SpringerBriefs in Physics

Hans László Pécseli

Introduction to the Theory
of Incoherent Scattering
# of Radar Waves
from Plasmas

 Springer

Hans László Pécseli ⓘ
Department of Physics
University of Oslo
Oslo, Norway

ISSN 2191-5423   ISSN 2191-5431 (electronic)
SpringerBriefs in Physics
ISBN 978-3-031-82651-1   ISBN 978-3-031-82652-8 (eBook)
https://doi.org/10.1007/978-3-031-82652-8

This Springer imprint is published by the registered company Springer Nature Switzerland AG
The registered company address is: Gewerbestrasse 11, 6330 Cham, Switzerland

If disposing of this product, please recycle the paper.

The book is dedicated to my family

Preface

The book presents an introduction to the theory of incoherent scattering of electromagnetic waves (radar waves) by free electrons in plasma media. The method is a standard tool for instance for monitoring important parameters of the Earth's ionosphere, but also for studies of hot plasmas such as in fusion experiments. The presentation begins with describing the scattering by statistically independent electrons, including magnetized cases. The model is then extended to account for "dressed particles" allowing for correlations between charged particles as induced by the long-range Coulomb forces.

The present notes are based on lectures for the course "Methods for investigating the near-Earth space environment" at the University of Tromso, Norway. The emphasis is here on the dressed particle model of fluctuations in plasmas, while the scattering process as such is covered in comparatively smaller detail. The present treatise attempts a presentation more like "Everyman's guide to incoherent scattering" with all the restrictions and simplifications that such an approach implies. Attention is given mostly to the collisionless kinetic plasma model but the book includes also suggestions for a collision model for charged particles with particular emphasis on the space-time evolution of the particles dress during collisions. The ideas are illustrated for ion collisions. This analysis is carried out in configuration space, and it is argued that the usual Fourier representation becomes incomprehensible in comparison.

The reader is assumed to be familiar with basic electro-magnetism, basic plasma physics, such as waves in plasma media, electrostatic waves in particular. It is an advantage to have some knowledge of kinetic plasma theory with some understanding of phenomena as described by the collisionless Vlasov equation. This information can be found in many textbooks on various levels of complexity, but the present appendices cover some of the essential points. The notes also contain a detailed reference list of the most relevant literature.

Oslo, Norway   Hans László Pécseli
November 2024

Acknowledgements I thank Jan Trulsen for many valuable discussions. Andres Spicher, Juha Vierinen, Ingrid Mann and Björn Gustavsson have all been giving constructive comments and encouragements during the writing of these notes. Patrick Guio contributed to much of the analysis presented in Appendix F.1.2 as found also in previous publications. In addition to published material I had access to unpublished lecture notes by Tor Hagfors, presented at the Nordic Research Course on *Radiation and scattering processes in Space Plasmas*, Sjudarhöjden, Sigtuna, Sweden, June 6–15, 1983, arranged by Uppsala Ionospheric Observatory.

Competing Interests The author has no competing interests to declare that are relevant to the content of this manuscript.

Contents

Chapter 1
Scattering by One Electron

Abstract As an introduction it is an advantage to study the scattering of an incoming electromagnetic wave by one free electron, and then "build-up" to a more general model. A classical approach suffices for this. For simplicity, only the unmagnetized case is discussed here.

1.1 Introduction

Incoherent scattering of electro-magnetic waves by free electrons (Thomson scattering) is a standard tool for monitoring the properties of the Earth's ionosphere and hot plasmas in general. The origin of the idea as such seems to be somewhat obscure, to be found in personal or private discussions. Charles Fabry [1] is cited for observing that free electrons in the ionosphere could scatter electromagnetic waves, in his case radiation from the sun. The first publications on scattering of radar beams seems to be by William Gordon [2] with experimental observations reported by Kenneth Bowles [3]. The ambition was to measure the electron density and electron temperature as a function of altitude and time at all levels in the Earth's ionosphere up to heights of one or more Earth's radii. The expectations were low due to the assumption that randomly distributed electrons would scatter individually. The cross section for the scattering by single electrons is small (as we will see) and the ideas were formulated late probably due to the fact that no radar installation was expected to have a sufficient power to give a detectable signal return. Sufficiently powerful radars were produced only a decade, or so, after the 2. World War. Incoherent scattering is one of the happy examples where reality turned out to be generous, with results significantly exceeding expectations. The reason is that electrons do not scatter individually, but are organized by collective interactions to become so called "dressed particles", giving significant increase in the available information. Many radar facilities have later been built for ionospheric research, EISCAT, Millstone Hill Observatory, Arecibo Observatory, Jicamarca Radio Observatory, Søndrestrøm Upper Atmospheric Research Facility, and Poker Flat Research Range, to mention some of the larger ones. Details can be found on the homepages for these facilities.

© The Author(s), under exclusive license to Springer Nature Switzerland AG 2025
H. L. Pécseli, *Introduction to the Theory of Incoherent Scattering of Radar Waves from Plasmas*, SpringerBriefs in Physics, https://doi.org/10.1007/978-3-031-82652-8_1

Incoherent scattering of electromagnetic waves is distinguished from Compton scattering. Following the standard terminology, the difference comes when dealing with atoms. In that case, if the scattering leaves the atom in the ground state, we deal with coherent scattering, whereas if the electron is ejected from the atom, the scattering is (incoherent) Compton scattering.

Scattering of electromagnetic waves by free electrons, Thomson scattering [4], is used as a remote diagnostic when the plasma is too warm to allow measurements carried out by probes immersed in the plasma [5, 6]. This will be the case in most plasma fusion experiments in Tokamaks and similar devices [7].

The following notes contain an extended version of lectures for the course FYS3002, "Methods for investigating the near-Earth space environment", at the University of Tromsø, presented February 2021, February 2022 and in part also in February 2023. Some preliminaries, here given as Appendices A–F, can be useful to make the presentation easier to follow. The present notes emphasize the dressed particle model for fluctuations in plasmas, while the scattering process as such is covered in comparatively smaller detail. The interested reader can find detailed and in-depth expositions in the literature [8–14]. The book includes a suggestion for a collision model for charged particles with particular emphasis on the space-time evolution of the particles dress during collisions. The ideas are illustrated for ion collisions. This analysis is carried out in configuration space, and it is argued that the usual Fourier representation becomes incomprehensible in comparison. There are some repitions in the text that hopefully makes it easier to read.

1.2   Electromagnetic Wave Scattering by One Electron

An incoming electromagnetic wave sets an electron into oscillatory motion. The accelerated electron will subsequently radiate another electromagnetic wave. By an observer, the process will be experienced as scattering of the incoming electromagnetic wave by the electron. The process can be adequately described within classical physics, as illustrated in the following. The scattered wave amplitude is small so "multiple scattering" can be ignored. Pioneering works are presented by several authors covering unmagnetized plasmas [15–18] as well as the magnetized case [19, 20]. A fine summary of relevant mathematical material can be found elsewhere [21].

Let the electric field of a linearly polarized incoming plane electromagnetic wave be $\mathbf{E}(\mathbf{r}, t) = \mathbf{E}_0 e^{-i(\omega_0 t - \mathbf{k}_0 \cdot \mathbf{r})}$. The wave scatters from a free electron with charge $-e$, initially at rest at a position $\mathbf{R}$ where the electric field is $\mathbf{E}(\mathbf{R}, t) = \mathbf{E}_0 e^{-i(\omega_0 t - \mathbf{k}_0 \cdot \mathbf{R})}$. The equation of motion for an unmagnetized electron is $m \, d\mathbf{u}/dt = -e \, \mathbf{E}_0 e^{-i(\omega_0 t - \mathbf{k}_0 \cdot \mathbf{R})}$ giving the oscillating velocity as

$$\mathbf{u}(t) = -i \frac{e \, \mathbf{E}_0 e^{-i(\omega_0 t - \mathbf{k}_0 \cdot \mathbf{R})}}{m \, \omega_0}, \tag{1.1}$$

with the electron charge and mass being $-e = -1.602 \times 10^{-19}$ Coulomb and $m = 9.110 \times 10^{-31}$ kg. The direction of the velocity vector $\mathbf{u}$ is parallel to $\mathbf{E}_0$. A magnetic force is here negligible in comparison to the electric forces, i.e., $|-e\mathbf{u} \times \mathbf{B}_0| \ll |-e\mathbf{E}_0|$, using $E_0/B_0 = c$ with $c \approx 299.792 \times 10^6$ ms^{-1} being the speed of light in vacuum. For relativistic conditions the magnetic force becomes important also [12], but for relevant ionospheric conditions we need not be concerned with these. The term $e^{i\mathbf{k}_0 \cdot \mathbf{R}}$ in (2.3) can here incorporated in $\mathbf{E}_0$. The oscillating position vector of the electron could obtained from (1.1) but it is not needed here. The maximum displacement of the electron is small for relevant electric fields so $\mathbf{R}$ is taken fixed in (1.1).

The magnitude of the net current associated with this oscillating electron is $\mathbf{j} = -e\mathbf{u}(t)\delta(\mathbf{r} - \mathbf{r}_e(t))$ where $\mathbf{r}_e(t)$ is the electron position. (Why is the physical dimension correct here? Readers unfamiliar with δ-functions can find support in Appendix A.) The vector potential at the receiver due to this current is given through

$$\mathbf{A}(\mathbf{r}, t) = \frac{\mu_0}{4\pi} \iiint_{-\infty}^{\infty} \frac{\mathbf{j}(\mathbf{r}', t')}{|\mathbf{r} - \mathbf{r}'|} d^3 r',$$

where $\mu_0 = 4\pi \times 10^{-7}$ Hm^{-1} is the magnetic permeability of free space. The current density $\mathbf{j}$ is to be obtained at a retarded time $t' = t - |\mathbf{r} - \mathbf{r}'_e|/c$ where $|\mathbf{r} - \mathbf{r}'_e|/c$ is the time it takes before a change is detected by the observer. The vector potential becomes here

$$\mathbf{A}(\mathbf{r}, t) = -\frac{e\mu_0}{4\pi} \frac{\mathbf{u}(t')}{|\mathbf{r} - \mathbf{r}'_e|}.$$

Being interested only in the "far field" of the radiation from the oscillating electron we may approximate as

$$\mathbf{A}(\mathbf{r}, t) \approx -\frac{e\mu_0}{4\pi} \frac{\mathbf{u}(t')}{|\mathbf{r}|} = -i \frac{\mu_0 e^2}{4\pi m \omega_0} \frac{\mathbf{E}_0 e^{-i\omega_0(t - |\mathbf{r}|/c)}}{|\mathbf{r}|}.$$

With $\omega_0 |\mathbf{r}|/c \equiv \mathbf{k}_1 \cdot \mathbf{r}$ the result is

$$\mathbf{A}(\mathbf{r}, t) \approx -i \frac{\mu_0 e^2}{4\pi m \omega_0} \frac{\mathbf{E}_0 e^{-i\omega_0 t + i\mathbf{k}_1 \cdot \mathbf{r}}}{R_1}, \tag{1.2}$$

where $R_1 = |\mathbf{r}|$ is the magnitude of the distance from detector (i.e. the observer) to the oscillating electron, see Fig. 1.1. The magnetic field associated with the radiation from the electron is then

$$\mathbf{B}(\mathbf{r}, t) = \nabla \times \mathbf{A}(\mathbf{r}, t) = \frac{\mu_0 e^2}{4\pi m \omega_0} \frac{\mathbf{k}_1 \times \mathbf{E}_0 \, e^{-i\omega_0 t}}{R_1} e^{i\mathbf{k}_1 \cdot \mathbf{r}}. \tag{1.3}$$

With $\mathbf{k}_1$ being along $\mathbf{R}_1$, see Fig. 1.1, we recognize $|\mathbf{E}_0 \times \mathbf{k}_1/k_1|$ as the magnitude of the incoming electric field component $\perp \mathbf{k}_1$. This field is linearly polarized in a plane containing the direction of $\mathbf{E}_0$ which is here also the direction of $\mathbf{u}(t)$.

Fig. 1.1 Schematic
illustration of the scattering
geometry. The position of
the transmitter (i.e. the radar)
is marked T, the
receiver/observer by B. The
volume containing the
scattering electron is marked
Vol. The transmitted radar
wave has wave-vector
$\mathbf{k}_0 \parallel \mathbf{R}_0$, the scattered wave
has $\mathbf{k}_1 \parallel \mathbf{R}_1$

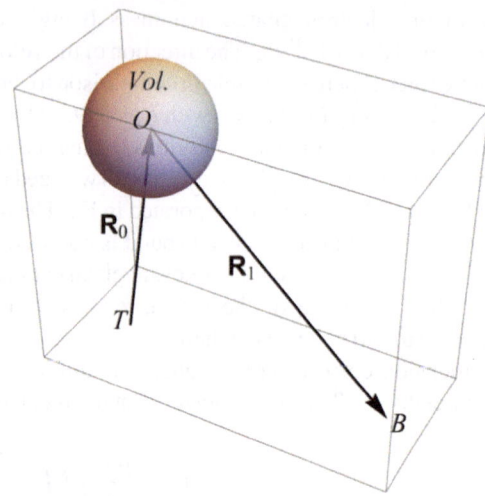

Use $c^2 \nabla \times \mathbf{B} = -i\omega_0 \mathbf{E}$ by Ampere's law with $\omega_0/k_1 = c$ and $c^2 = (\varepsilon_0 \mu_0)^{-1}$ to find the electric field of the scattered wave

$$\mathbf{E}(\mathbf{r}, t) = -\frac{e^2}{4\pi \varepsilon_0 m \omega_0^2} \frac{\mathbf{k}_1 \times (\mathbf{k}_1 \times \mathbf{E}_0)e^{-i\omega_0 t}}{R_1} e^{i\mathbf{k}_1 \cdot \mathbf{r}}$$

$$= -\left(\frac{r_0}{R_1}\right) \hat{\mathbf{q}} \times (\hat{\mathbf{q}} \times \mathbf{E}_0)e^{-i(\omega t - \mathbf{k}_1 \cdot \mathbf{r})}. \tag{1.4}$$

The permittivity of free space $\varepsilon_0 \approx 8.854 \times 10^{-12}$ Fm^{-1} enters and $\hat{\mathbf{q}} \equiv \mathbf{k}_1/k_1$ is the unit vector in the direction of propagation of the scattered radiation from the scattering electron to the observer, see Fig. 1.1. The classical electron radius introduced here is $r_0 \equiv e^2/(4\pi \varepsilon_0 m c^2) = 2.82 \times 10^{-15}$ m. It can be interpreted as the distance between two electrons when their relative potential energy is mc^2.

The magnitude of the scattered Pointing flux $\mathbf{E} \times \mathbf{B}/\mu_0$ is then

$$S_r = \frac{1}{2}\sqrt{\frac{\mu_0}{\varepsilon_0}} \left(\frac{e^2}{4\pi m \omega_0}\right)^2 \frac{|\mathbf{k}_1 \times \mathbf{E}_0|^2}{R_1^2},$$

We have $|\mathbf{k}_1| = |\mathbf{k}_0|$ for this particular simple case. The radiation originating from the accelerated electron is experienced as scattering of the incoming (forcing) electromagnetic wave.

The scattering angle χ is introduced through

$$\sin \chi \equiv \frac{|\mathbf{k}_1 \times \mathbf{E}_0|}{|\mathbf{k}_1||\mathbf{E}_0|}.$$

Two ω_0's cancel when k_1 is inserted to give the result

$$S_r = \frac{3}{2}\sigma_T \sin^2\chi \frac{S_{in}}{4\pi R_1^2}, \qquad (1.5)$$

in terms of the incoming Pointing flux $S_{in} = \frac{1}{2}|E_0|^2\sqrt{\varepsilon_0/\mu_0}$ and the Thomson cross section $\sigma_T \equiv \frac{8}{3}\pi r_0^2 = 6.6 \times 10^{-29}$ m^2. The χ–dependence of the radiation pattern can be understood as a pattern from a small dipole antenna, i.e., the electron oscillating along the electric field direction. The radiation pattern is illustrated in Fig. 1.2. There is no radiation observed strictly in the direction of the electron motion, i.e., when $\mathbf{k}_1 \parallel \mathbf{E}_0$. The presence of the "plane-wave" exponential $e^{i\mathbf{k}_1\cdot\mathbf{r}}$ in the expressions for $\mathbf{E}$ and $\mathbf{B}$ can be understood by noting that in the far field (at a long distance from the oscillating electron, the distant or "Fraunhofer" field) the waveform can be seen as a locally plane wave. This will be true also for spherical waves, just as for the radiation pattern shown in Fig. 1.2.

As an illustration we can imagine an electromagnetic wave-packet propagating in a homogeneous plasma with density n. The mean free path for scattering $(n\sigma_T)^{-1}$ is of the order of kilometers for a plasma with density $n \sim 10^{25}$ m^{-3}. This will be a large density for most laboratory plasmas so it can be concluded that multiple scattering is highly unlikely there [22]. Long mean-free paths in the sense outlined here can be found by scattering from the ionosphere, for instance. The previous estimate also shows that very energetic sources of electromagnetic waves are required to prevent the scattered signal being swamped by spontaneous radiation from the plasma. For ionospheric applications, the signal has to be sufficiently intense to be measurable by detectors on ground.

The result (1.5) can be expressed in a different form, sometimes found in the literature [22], as the energy scattered per second per unit solid angle $d\Omega$

$$\frac{dW}{dt\,d\Omega} = S_{in}r_0^2 \left| \hat{\mathbf{q}} \times \left(\hat{\mathbf{q}} \times \frac{\mathbf{E}_0}{E_0} \right) \right|^2 .$$

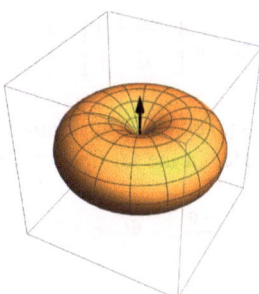

Fig. 1.2 Illustration of the radiation pattern from a single electron for varying directions as determined by χ. The incoming electromagnetic wave is assumed to be linearly polarized with the electric field vertical. The angle between $\mathbf{E}_0$ and the direction vector to the observer is χ

The total power scattered by an electron is given by

$$P_T = \int S_r R^2 d\Omega = \frac{3}{2}\sigma_T \frac{S_{in}}{4\pi} \int \sin^2\chi \, d\Omega = S_{in}\sigma_T$$

where $d\Omega$ is the angular element. The Thomson cross section is therefore the ratio between the total power scattered by one electron and the incident power density.

So far the electric field was assumed to be linearly polarized and the reference direction to be along the field. The angle χ is then the angle between the direction to the observer and $\mathbf{E}_0$. The full problem, i.e. the general relation between the electric field direction, and the directions to the emitter and the receiver is more complicated [22]. Consider a harmonic wave propagating in the z-direction

$$\mathbf{E}_0 = \mathcal{E}_0 \left(\cos\beta\,\hat{\mathbf{x}} + e^{i\delta}\sin\beta\,\hat{\mathbf{y}} \right) \equiv \mathcal{E}_0 \left\{ \begin{matrix} \cos\beta \\ e^{i\delta}\sin\beta \end{matrix} \right\} \equiv \mathcal{E}_0\,\mathbf{p}. \qquad (1.6)$$

Linear polarization along the y-axis corresponds to $\mathbf{p} = \left\{ \begin{matrix} 1 \\ 0 \end{matrix} \right\}$ and along the x-axis to $\mathbf{p} = \left\{ \begin{matrix} 0 \\ 1 \end{matrix} \right\}$. Right circular polarization is the obtained by $\mathbf{p} = \frac{1}{\sqrt{2}} \left\{ \begin{matrix} 1 \\ i \end{matrix} \right\}$, while left circular polarization is found for $\mathbf{p} = \frac{1}{\sqrt{2}} \left\{ \begin{matrix} 1 \\ -i \end{matrix} \right\}$.

In order to specify the polarization both of the transmitted and the scattered wave the following coordinates are introduced for the incoming (or transmitted) wave $\hat{\mathbf{z}} \equiv \mathbf{k}_0/|\mathbf{k}_0|$, together with $\hat{\mathbf{y}}$ being a vector normal to the plane defined by $\mathbf{k}_0$ and $\mathbf{k}_1$, while $\hat{\mathbf{x}}$ is normal to both $\hat{\mathbf{z}}$ and $\hat{\mathbf{y}}$ in a right handed system. For the scattered wave we have $\hat{\mathbf{z}}' \equiv \hat{\mathbf{q}} = \mathbf{k}_1/|\mathbf{k}_1|$, $\hat{\mathbf{y}}' = \hat{\mathbf{y}}$ and $\hat{\mathbf{x}}'$ being normal to both $\hat{\mathbf{z}}'$ and $\hat{\mathbf{y}}'$ in a right handed system. In terms of these coordinate vectors we can write

$$\left\{ \begin{matrix} E'_{rx} \\ E'_{ry} \end{matrix} \right\} = \frac{r_0}{R_1}\mathcal{E}_0 \left\{ \begin{matrix} \cos\theta & 0 \\ 0 & 1 \end{matrix} \right\} \cdot \left\{ \begin{matrix} \cos\beta \\ e^{i\delta}\sin\beta \end{matrix} \right\} e^{-i(\omega_0 t - \mathbf{k}_1\cdot\mathbf{r})}, \qquad (1.7)$$

or

$$\mathcal{E}_{r0} \left\{ \begin{matrix} \cos\beta' \\ e^{i\delta'}\sin\beta' \end{matrix} \right\} = \frac{r_0}{R_1}\mathcal{E}_0 \left\{ \begin{matrix} \cos\theta & 0 \\ 0 & 1 \end{matrix} \right\} \cdot \left\{ \begin{matrix} \cos\beta \\ e^{i\delta}\sin\beta \end{matrix} \right\} e^{-i(\omega_0 t - \mathbf{k}_1\cdot\mathbf{r})}. \qquad (1.8)$$

Note that $\sin^2\chi = \sin^2\beta + \cos^2\beta\cos^2\theta$. The relation between the scattered and the incoming electric field amplitudes in the far field can be approximated by

$$\mathcal{E}_{r0}\,\mathbf{p}' = \frac{r_0}{R_1}\,\mathcal{E}_0\,\underline{\boldsymbol{\Psi}}\cdot\mathbf{p}\,e^{-i(\omega_0 t - \mathbf{k}_1\cdot\mathbf{r})}, \qquad (1.9)$$

where

$$\underline{\boldsymbol{\Psi}} \equiv \left\{ \begin{matrix} \cos\theta & 0 \\ 0 & 1 \end{matrix} \right\} \qquad \text{a matrix accounting for the geometry}$$

$$\mathbf{p} \equiv \left\{ \begin{array}{c} \cos \beta \\ e^{i\delta} \sin \beta \end{array} \right\} \qquad \text{a vector accounting for the transmitter}$$

$$\mathbf{p}' \equiv \left\{ \begin{array}{c} \cos \beta' \\ e^{i\delta'} \sin \beta' \end{array} \right\} \qquad \text{a vector accounting for the receiver}$$

The interpretation of the matrix-vector product implied in $\underline{\mathbf{\Psi}} \cdot \mathbf{p}$ is made evident in (1.7) and (1.8). The description summarized in this latter part of the section is particularly useful for circularly or elliptically polarized waves.

Chapter 2
Scattering by Many Electrons

Abstract Given the results of electromagnetic waves scattered by one electron it is possible to give a simple generalization for many independent electrons, allowing also for their thermal or random motions. The emphasis is on the unmagnetized case, but a summary of scattering from magnetized plasmas is included as well.

2.1 Scattering by Many Non-interacting Electrons

Assume that the plasma electrons are randomly and independently distributed in an infinite space. Can it be so that the scattering from one will always be cancelled by scattering from some other one? This is not the case. Sometimes the signals add, sometimes they subtract in such a way that they cancel on average, but the root mean square contribution is finite. This problem is illustrated by the simple model signal described in Appendix B.1.2 and elaborated further in what follows. The terms "incoherent scattering" refers to scattering objects (it *need* not be free individual electrons) randomly and independently distributed. X-ray scattering from crystals represents the other extreme with ordered scatterers.

If all electrons oscillate with respect to a fixed reference point for each, they will return oscillations with the same frequency as the one that sets them into motion: the plasma will act as a mirror, a rather poor one as such. If the electron reference point is moving during the electron oscillation, we will observe an broadening of this return frequency line due to the resulting Doppler shifts. Details of this back-scatter spectrum contains information of some of the plasma parameters. The objective of the following analysis is to give a physical model for this line broadening.

The geometry of the present problem is illustrated in Fig. 2.1. The incoming electric field at the position of a selected electron j on the figure can be written as

$$\mathbf{E}_i = \mathbf{E}_0 e^{-i(\omega_0 t - \mathbf{k}_0 \cdot \mathbf{R}_p)}.$$

This electron gives rise to a scattered electric field at the detector (receiver) position

$$E_{rj} = \mathcal{E}_0 e^{-i(\omega_0 t - \mathbf{k}_0 \cdot \mathbf{R}_p)} e^{i \mathbf{k}_1 \cdot \mathbf{R}_j},$$

© The Author(s), under exclusive license to Springer Nature Switzerland AG 2025
H. L. Pécseli, *Introduction to the Theory of Incoherent Scattering of Radar Waves from Plasmas*, SpringerBriefs in Physics, https://doi.org/10.1007/978-3-031-82652-8_2

Fig. 2.1 Schematic illustration of the position of a reference electron labelled j, marked by a small black disk at position R. The position of the transmitter (i.e. the radar) is marked T, the receiver/observer by B. The volume containing the scattering electrons is marked *Vol*

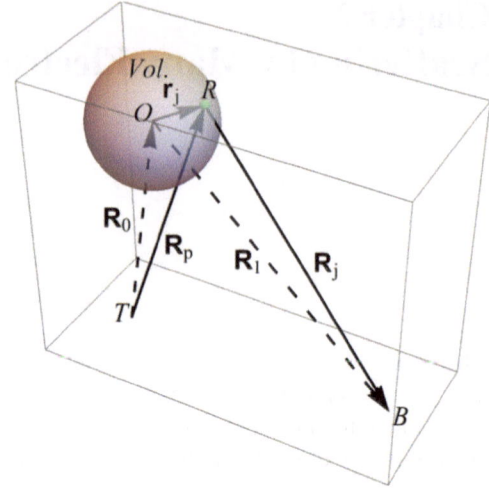

where some constant factors are included in $\mathcal{E}_0$. The distance between the receiver and the center of the entire scattering volume where the electrons are distributed is taken to be R_1. The position of an electron with respect to this central position marked O in Fig. 2.1 is taken to be $\mathbf{r}_j$. Referring to Fig. 2.1 illustrating the relations $\mathbf{R}_p = \mathbf{R}_0 + \mathbf{r}_j$ and $\mathbf{R}_1 = \mathbf{R}_j + \mathbf{r}_j$ we can write

$$E_{rj} = \mathcal{E}_0 e^{-i(\omega_0 t - \mathbf{k}_0 \cdot (\mathbf{R}_0 + \mathbf{r}_j))} e^{i\mathbf{k}_1 \cdot (\mathbf{R}_1 - \mathbf{r}_j)}.$$

Assume $r_j \ll R_1, R_0$ for all j so that for instance $1/|\mathbf{r}_j + \mathbf{R}_1| \approx 1/|\mathbf{R}_1|$. Implicitly the lines $B - O$ and $B - R$ are taken to be parallel, and similarly for the lines $T - O$ and $T - R$. The distances $|\mathbf{R}_0|$ and $|\mathbf{R}_1|$ are thus assumed to be large, so that the directions of the vectors change negligibly when moving from one scattering electron to another inside reference the volume in Fig. 2.1.

The previous expression can be simplified by introducing

$$\mathbf{k} = \mathbf{k}_0 - \mathbf{k}_1. \tag{2.1}$$

Sometimes, in particular in solid state physics, the wavevector $\mathbf{k}$ in (2.1) is denoted the *Bragg vector* $\mathbf{k}_B$. To illustrate the concept of the Bragg vector you might imagine a small mirror placed in the overlapping region of a transmitting and receiving radar, see Fig. 2.1. Only one orientation of the mirror is optimum for reception in this particular case, and the Bragg vector is here perpendicular to the mirror surface.

The magnitude of the electric field E_r resulting from scattering by N electrons at positions distributed in a volume V centered at a position $\mathbf{R}_1$ can now be written as the sum of many independent contributions

$$E_r(t) = \mathcal{E}_0 e^{-i(\omega_0 t - \mathbf{k}_0 \cdot \mathbf{R}_0)} e^{i\mathbf{k}_1 \cdot \mathbf{R}_1} \sum_{j=1}^{N} e^{i\mathbf{k} \cdot \mathbf{r}_j}. \tag{2.2}$$

The complex coefficient $\mathcal{E}_0 e^{-i(\omega_0 t - \mathbf{k}_0 \cdot \mathbf{R}_0)} e^{i\mathbf{k}_1 \cdot \mathbf{R}_1}$ is in reality nothing but the scatter from one electron placed at O, so the sum in (2.2) is the correction that accounts for the contribution of all N electrons. Previously the scatter from one electron was found to be expressed as $\mathbf{E}_r = (r_0/R_1)\mathcal{E}_0 \underline{\boldsymbol{\Psi}} \cdot \mathbf{p}\, e^{-i(\omega_0 t - \mathbf{k}_1 \cdot \mathbf{r})}$. The results (2.2) can therefore be written as

$$\mathbf{E}_r = \frac{r_0}{R_1} \mathcal{E}_0 \underline{\boldsymbol{\Psi}} \cdot \mathbf{p}\, e^{-i(\omega_0 t - \mathbf{k}_1 \cdot \mathbf{r})} \sum_{j=1}^{N} e^{i\mathbf{k}\cdot\mathbf{r}_j}.$$

To express this result in terms of the electron density $n(\mathbf{r}, t)$ we write

$$n(\mathbf{r}, t) = \sum_{k} n(\mathbf{k}, t) e^{-i\mathbf{k}\cdot\mathbf{r}} \qquad \text{with} \qquad n(\mathbf{k}, t) = \iiint_V n(\mathbf{r}, t) e^{i\mathbf{k}\cdot\mathbf{r}} d^3 r.$$

A time variation of the density is allowed for by the motion of particles, $\mathbf{r}_j = \mathbf{r}_j(t)$, as discussed in the following. Since the electron density can be expressed as $n(\mathbf{r}, t) = \frac{1}{V}\sum_{j=1}^{N} \delta\left(\mathbf{r} - \mathbf{r}_j(t)\right)$, we have

$$n(\mathbf{k}, t) = \frac{1}{V} \sum_{j=1}^{N} e^{i\mathbf{k}\cdot\mathbf{r}_j} = \frac{1}{V} \sum_{j=1}^{N} n_j(\mathbf{k}, t), \tag{2.3}$$

giving

$$E_r(t) = \mathcal{E} V n(\mathbf{k}, t), \tag{2.4}$$

where a generally complex quantity $\mathcal{E}$ was introduced as a shorthand for the magnitude of the coefficient of the sum in (2.2). The electric field direction is assumed to be the same at all electron positions, so there is no need to sum all $\underline{\boldsymbol{\Psi}} \cdot \mathbf{p}$'s.

The relation (2.4) states that the radar beam is in effect Fourier transforming the spatially varying plasma density [21]. This is an important result! Note that $\mathbf{k}$ is not a free variable: if you want another $\mathbf{k}$ you have to change the radar frequency or change the geometry, e.g., by moving the receiver. If it so happens that $n(\mathbf{k}, t)$ is small you get a small reflection although there might be plenty of electrons.

A statistical average of E_r is vanishing because of the mixing of all the phases. To get a useful result you have to "square" the result somehow. The correlation function (see Sect. B.1.4) is good candidate since it contains more information than a simple $\langle |E_r|^2 \rangle$. An expression for the time varying auto correlation function can be obtained by

$$\langle E_r(t) E_r^*(t + \tau) \rangle = |\mathcal{E}|^2 e^{i\omega_0 \tau} n_0 V \left\langle n_j(\mathbf{k}, t) n_j^*(\mathbf{k}, t + \tau) \right\rangle, \tag{2.5}$$

with $n_0 V = N$. The $e^{i(\mathbf{k}_0 \cdot \mathbf{R}_0 + \mathbf{k}_1 \cdot \mathbf{R}_1)}$-term in (2.2) vanishes when taking the absolute value, while an $e^{i\omega_0 \tau}$-term remains.

For the present problem with (2.3) the (2.5) auto-correlation function is found as

$$\left\langle n_j(\mathbf{k}, t) n_j^*(\mathbf{k}, t + \tau) \right\rangle = \left\langle e^{i\mathbf{k}\cdot\mathbf{r}_j(t) - i\mathbf{k}\cdot\mathbf{r}_j(t+\tau)} \right\rangle \equiv \rho(\mathbf{k}, \tau).$$

The τ variation vanishes when the electron oscillation centers $\mathbf{r}_j$ are fixed and then $\rho(\mathbf{k}, \tau) = 1$. A less trivial result is found when the electron oscillation center moves during the time interval τ due to its thermal motion, for instance. A major part of the following text is concerned with determining an expression for $\rho(\mathbf{k}, \tau)$.

2.1.1 Moving Electrons

Assume now that the unperturbed electrons are moving without collisions along straight-line orbits, $\mathbf{r} = \mathbf{r}_0 + \mathbf{u}t$. Relevant particle velocities, $\mathbf{u}$, are small in the sense that an electron moves only little during a period of oscillation. The electron velocities are assumed to have a known distribution, for example a Maxwellian $f_0(\mathbf{u}) = (2\pi u_{th}^2)^{-3/2} e^{-u^2/2u_{th}^2}$. The present straight line orbits assume collisionless plasmas. In reality the charged particles are interacting and their assumed straight line orbits can be perturbed. We make the assumption here anyhow, and discuss the inconsistencies later.

A general and useful analytical result is found in the form

$$\left\langle \sum_{j=1}^N G(\mathbf{u}_j) \right\rangle = n_0 V \iiint G(\mathbf{u}) f_0(\mathbf{u}) d^3 u, \qquad (2.6)$$

for any integrable function G, and with large N, where again $n_0 V = N$ and $f_0(\mathbf{u})$ is the normalized particle velocity distribution, $\iiint f_0(\mathbf{u}) d^3 u = 1$, not necessarily a Maxwellian. The large number N of particles are assumed to be uniformly distributed in the volume V. As a simple test it is found that (2.6) is trivially satisfied for $G = 1$.

With the assumed Maxwellian distribution we find

$$\rho(\mathbf{k}, \tau) = \left\langle e^{i\mathbf{k}\cdot(\mathbf{r}_{0j}+\mathbf{u}_j t)-i\mathbf{k}\cdot(\mathbf{r}_{0j}+\mathbf{u}_j(t+\tau))} \right\rangle = e^{-(ku_{th}\tau)^2/2}, \qquad (2.7)$$

with a correlation normalized at the origin, $\rho(\mathbf{k}, 0) = 1$. Multiply by $n_0 V$ to get the un-normalized version. In (2.7) the time-correlation of the return-signal contains a measure of the thermal velocity. Already here it is found that incoherent scattering has the capacity of giving information of the electron density and, by a fit to the correlation function, also the electron temperature. The electron velocity distribution $f_0(\mathbf{u})$ can be generalized to allow for a bulk relative plasma motion, i.e., a current, but the correlation function is noticeably complicated by this, in part because a new reference vector (the velocity) is introduced in addition to $\mathbf{k}$.

Rather than the correlation function, it is often preferable to use the power spectrum, which can be defined as the Fourier transform of the correlation function by the Wiener-Khinchin theorem [21–23]. For this particular case we have

$$G(\mathbf{k}, \omega) = \int_{-\infty}^{\infty} \rho(\mathbf{k}, \tau) e^{-i\omega\tau} d\tau = e^{-\frac{1}{2}(\omega/ku_{th})^2} \sqrt{\frac{2\pi}{(ku_{th})^2}}. \qquad (2.8)$$

You might find a definition differing by a factor 2 due to a single sided transform. The frequency ω is the correction to the radar frequency ω_0 which appears by the $e^{i\omega_0\tau}$ coefficient remaining in (2.5): this factor introduces a frequency shift by the Fourier transform [21]. Spectral components in the Fourier transform of the correlation function will therefore be side-bands to the carrier frequency ω_0. For given k, the frequency spectrum narrows to a δ-function when $u_{th} \to 0$, see Appendix A. In that limit the electrons centers of oscillation are at rest, the correlation function is unity and the electrons merely reflect the incoming radar beam with frequency ω_0.

2.2 Magnetized Plasmas

For a magnetized plasma the general particle orbit has a helical form along **B**, see Fig. 2.2. The velocity magnitude (the speed) is constant along the obit for homogeneous magnetic fields, but we have to account for the changing direction of the velocity vector when expressing $\mathbf{r} = \mathbf{r}_0 + \mathbf{u}t$. Although magnetic fields found in space are inhomogeneous on large scales, the analytical studies are almost universally assuming locally homogeneous fields.

Assume that a homogeneous magnetic field is directed along the z-axis. It is convenient to operate with two different coordinate systems, a Cartesian and a set of polarized coordinates to describe the motion of a magnetized charged particle

$$\text{Cartesian}: \quad \mathbf{r} = \{x, y, z\}, \quad \text{and polarized}: \quad \mathbf{r} = \{r_{+1}, r_{-1}, r_0\}$$

The two coordinate systems are related by $r_{+1} = \frac{1}{\sqrt{2}}(x + iy)$, $r_{-1} = \frac{1}{\sqrt{2}}(x - iy)$ and $r_0 = z$. Making advantage of the polarized coordinates, the free motion of the magnetized particles is then expressed as

$$\frac{du_\alpha}{dt} = -i\alpha\Omega_c u_\alpha, \quad \alpha = (\pm 1, 0), \quad \Omega_c = \frac{qB_0}{m},$$

with q, m being the charge and mass of the particle (usually singly charged ions or electrons). The cyclotron frequency is Ω_c with the sign of the charge included. The velocity at time $t' = t - \tau$ is related to the velocity at t through

$$\mathbf{u}(t - \tau) = \underline{\dot{\Gamma}}(\tau) \cdot \mathbf{u}(t),$$

Fig. 2.2 The spiral orbit followed by a charged particle moving in a homogeneous magnetic field

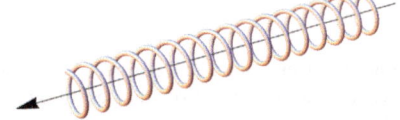

where $\underline{\dot{\Gamma}}(\tau) \equiv d\underline{\Gamma}/d\tau$ is a diagonal matrix with elements $\dot{\Gamma}_{\alpha\alpha}(\tau) = e^{i\alpha\Omega_c\tau} \equiv \dot{g}_\alpha(\tau)$. The past positions can be expressed in terms of the present position by $\mathbf{r}(t - \tau) = \mathbf{r}(t) - \underline{\Gamma}(\tau) \cdot \mathbf{u}(t)$. The helical particle trajectory is determined by $\underline{\Gamma}$ with elements

$$\Gamma_{\alpha\alpha}(\tau) = \frac{e^{i\alpha\Omega_c\tau} - 1}{i\alpha\Omega_c} = g_\alpha(\tau), \quad \alpha = (\pm 1, 0),$$

not to be confused with the usual Γ-function.

The auto correlation function becomes

$$\rho_p(\mathbf{k}, \tau) = \langle e^{i\mathbf{a}\cdot\mathbf{u}}\rangle, \tag{2.9}$$

with vector components $a_\alpha(\tau) = k_\alpha g_\alpha(\tau)$. The form of the correlation function depends on the particle velocity distribution, the strength of the magnetic field and the angle θ between $\mathbf{k}$ and $\mathbf{B}_0$. Assuming a Maxwellian velocity distribution

$$f_0(\mathbf{u}) = \frac{e^{-u^2/2u_{th}^2}}{(2\pi u_{th}^2)^{3/2}} \quad \text{with} \quad u^2 = u_x^2 + u_y^2 + u_z^2,$$

we find the auto-correlation function corresponding to (2.7) to be

$$\rho(k, \theta, \tau) = e^{-\frac{1}{2}|\mathbf{a}(\tau)|^2 u_{th}^2} = e^{-(kr_L)^2\left((\Omega_c\tau/2)^2\cos^2\theta + \sin^2(\Omega_c\tau/2)\sin^2\theta\right)}, \tag{2.10}$$

where $r_L = \sqrt{2}u_{th}/\Omega_c$ is the electron Larmor radius. The components of the vector $\mathbf{a}$ were defined in relation to (2.9). See Fig. 2.3 for illustrations of (2.10). The polarization of the electric field is no longer trivial since there is now a preferred direction given by the magnetic field. The direction of rotation for circularly (or elliptically) polarized fields becomes important too; a reference circulation is given by the direction of the gyro-rotation of the electrons.

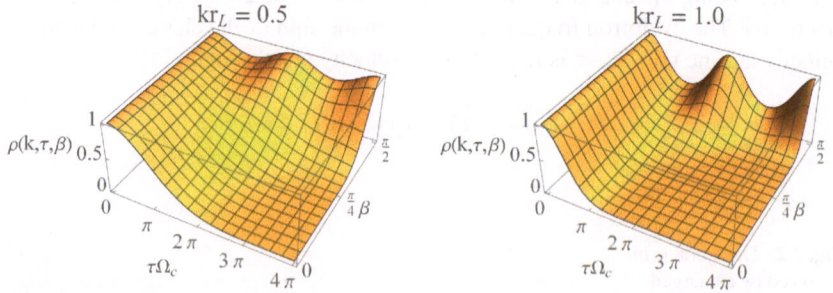

Fig. 2.3 Correlation functions for statistically independent, non-interacting scatterers in magnetized plasmas for two choices of kr_L. The figure assumes linearly polarized incoming electromagnetic waves. For $\theta = 0$, the electric field is along $\mathbf{B}$

Chapter 3
Dressed Particles

Abstract It turns out that the model of electro-magnetic waves scattered by randomly distributed individual electrons is oversimplified. The long range Coulomb forces give rise to a correlation of the electrons and the can not be assumed of the scattering individually. The result is a significant modification of scattering process. This long range ordering can be accounted for by a "dress" moving with the electrons. Ions are too heavy to participate in the scattering, but by structuring the surrounding electron cloud they will contribute indirectly to the scattering.

3.1 Dressed Particles

A moving electron gives rise to a polarization of its surroundings. For a stationary charge, the perturbation of the surroundings will be given by the Debye cloud, giving the shielding as discussed in Appendix C. For velocities small compared to the thermal velocity, the perturbation has the form of a moving and slightly distorted Debye shielding. Super-thermal electrons will, on the other hand be followed by a radiation pattern composed of Langmuir waves, in a manner somewhat similar to the waves following a moving ship. These perturbations will move together with the reference particle. The physics of "dressed particles" is basically the same for any moving object in matter that can support wave-like motion, see for instance Fig. 3.1. The dressed electrons are still taken to be randomly and independently distributed, so in this sense the scattering is still incoherent. The concept and in particular also illustrations of dressed particles is discussed in more detail in Appendix F.

3.1.1 Dressed Electrons

All the electrons in the plasma can be taken one by one, letting their interaction with all the others being represented by an associated "dress". These dressed particles are moving with statistically distributed velocities, following e.g. a Maxwellian

© The Author(s), under exclusive license to Springer Nature Switzerland AG 2025   15
H. L. Pécseli, *Introduction to the Theory of Incoherent Scattering of Radar Waves from Plasmas*, SpringerBriefs in Physics, https://doi.org/10.1007/978-3-031-82652-8_3

Fig. 3.1 The physics of "dressed particles" is basically the same for any object moving in matter that can support waves, here surface waves on water. You might imagine many independent ducks swimming around with their wake fields adding up in the present linear model. Each small parcel of water participates in the linearly superimposed wave pattern generated by many different ducks. The photo is reproduced with the kind permission of Professor Dietrich Zawischa.

distribution. The "dresses" are linearly added, nonlinear interactions are not included. In a fluid model the "dress" is the Coulomb shielding (possibly with some distortions) for subthermal particles, while for superthermal particles it is the radiation pattern illustrated in Fig. F.1. This dressed particle approach was seemingly first suggested by Thompson [24] and later elaborated by many authors [15, 18, 25, 26]. A review of some elements of the analysis are given also in textbooks [22, 27, 28]. To account for the dress of the reference particle we need to be concerned only with longitudinal or electrostatic waves. Also transverse wave will be excited in plasmas [22] but these will not give rise to density fluctuations and are not relevant for incoherent scattering.

It is the light charged particles, i.e. electrons, that do the scattering. Ions contribute indirectly: a fluctuating ion density induces a corresponding electron perturbation to maintain a local, approximate, quasi-neutrality, see Appendix C.1.9. Thereafter, the scattering will contain the signature of the ion density variation.

First the dielectric function $\varepsilon_r(\mathbf{k}, \omega)$ of the plasma is assumed to be known. Taking again an electron moving along a straight line trajectory as the reference "external" particle labeled j, the associated charge density will be $-en_j(\mathbf{r}, t) = -e\delta(\mathbf{r} - \mathbf{r}_{0j} - \mathbf{u}_j t)$. Its Fourier transform is

$$- en_j(\mathbf{k}, \omega) = -2\pi e\delta(\omega t - \mathbf{k} \cdot \mathbf{u}_j)e^{i\mathbf{k} \cdot \mathbf{r}_{0j}}. \tag{3.1}$$

The Fourier transform of the induced potential is then

$$\phi(\mathbf{k}, \omega) = -e\frac{2\pi\delta(\omega - \mathbf{k} \cdot \mathbf{u})e^{i\mathbf{k} \cdot \mathbf{r}_0}}{k^2\varepsilon_0\varepsilon_r(\mathbf{k}, \omega)},$$

where $\mathbf{r}_0$ is the initial particle position. By removing one electron the composition of the plasma and thereby also its dielectric function is changed. Since, however, it was assumed that there are many electrons in a Debye-sphere (large plasma parameters N_p, see Appendix C.1.5), the changes are insignificant and the original dielectric function is retained. For small plasma parameters it will on the other hand be permissible to ignore particle interactions altogether and use the previous models for individually moving single electrons.

It is, however, not the electrostatic potential that is needed but the electron density. Using (D.4) with $\zeta = -e n_{int}$, the charges e will cancel to give

$$n_{int}(\mathbf{k}, \omega) = (1 - \varepsilon_r(\mathbf{k}, \omega)) \frac{2\pi \delta(\omega - \mathbf{k} \cdot \mathbf{u}) e^{i\mathbf{k} \cdot \mathbf{r}_0}}{\varepsilon_r(\mathbf{k}, \omega)},$$

being the plasma density variation induced by the selected reference electron. Integrating the charge density $-e n_{int}(\mathbf{r}, t)$ associated with the particle dress over all space we find a net charge that exactly compensates the selected reference charge, just as in the case of the Debye shielding of stationary charges, see Appendix C.1.

The Fourier transform of the entire induced density variation is found by taking all the electrons in the reference volume, one by one, adding their contributions up to give

$$n_{int}(\mathbf{k}, \omega) = 2\pi \frac{(1 - \varepsilon_r(\mathbf{k}, \omega))}{\varepsilon_r(\mathbf{k}, \omega)} \sum_{j=1}^{N} \delta(\omega - \mathbf{k} \cdot \mathbf{u}_j) e^{i\mathbf{k} \cdot \mathbf{r}_{0j}}.$$

Using (3.1), the reference electrons alone are accounted for by the expression

$$n_{ext}(\mathbf{k}, \omega) = 2\pi \sum_{j=1}^{N} \delta(\omega - \mathbf{k} \cdot \mathbf{u}_j) e^{i\mathbf{k} \cdot \mathbf{r}_{0j}}.$$

The Fourier transform of the total electron density, the moving reference electrons and their dress, $n(\mathbf{k}, \omega) = n_{ext}(\mathbf{k}, \omega) + n_{int}(\mathbf{k}, \omega)$, is then

$$n(\mathbf{k}, \omega) = 2\pi \sum_{j=1}^{N} \delta(\omega - \mathbf{k} \cdot \mathbf{u}_j) e^{i\mathbf{k} \cdot \mathbf{r}_{0j}} \left(1 - \frac{\varepsilon_r(\mathbf{k}, \omega) - 1}{\varepsilon_r(\mathbf{k}, \omega)} \right). \tag{3.2}$$

This result is often expressed in terms of the susceptibility $\chi(\mathbf{k}, \omega) = \varepsilon_r(\mathbf{k}, \omega) - 1$.

In the large parenthesis of (3.2) the "unity" accounts for the reference electron that causes the screening or "dress", the denominator accounts for the electrostatic potential induced by the moving charge while the nominator transforms the potential to the desired electron density. Be aware that if you could follow a moving test charge you would hardly notice the dress, just as for the Debye cloud discussed in Appendix C.1.6. You would have to follow the reference particle for a long time and take the average over its surroundings to get a visualization of the "dress".

The analysis so far assumed immobile ions. In reality the thermal ion motion has a significant effect on the scattering, so ways to include also this has to be found. This problem is addressed in the following section.

3.1.2   *Inclusion of Ion Motion*

Ions are too heavy to be put into any significant motion of the high frequency electric field from a radar, but the ion component can affect the electron density indirectly [16]. A moving ion will attract nearby electrons and be surrounded by a slight electron surplus. Thus also moving ions have a dress following them. The local electron density will be influenced and thereby also the electron scattering. The ion dress is found in very much the same way as for the electrons: first the electrostatic potential following a moving ion is determined using this to obtain the *electron* density. Since the ions do not scatter, the ion density associated with the dress is not needed but it could easily found too.

The result for the Fourier transform of the total electron density then becomes

$$n(\mathbf{k}, \omega) = 2\pi \sum_{j=1}^{N} \delta(\omega - \mathbf{k} \cdot \mathbf{u}_j^{(e)}) e^{i\mathbf{k} \cdot \mathbf{r}_{0j}} \left(1 - \frac{\varepsilon_r^{(e)}(\mathbf{k}, \omega) - 1}{\varepsilon_r^{(e)}(\mathbf{k}, \omega) + \varepsilon_r^{(i)}(\mathbf{k}, \omega) - 1} \right) +$$

$$2\pi \sum_{j=1}^{N} \delta(\omega - \mathbf{k} \cdot \mathbf{u}_j^{(i)}) e^{i\mathbf{k} \cdot \mathbf{r}_{0j}} \left(\frac{\varepsilon_r^{(e)}(\mathbf{k}, \omega) - 1}{\varepsilon_r^{(e)}(\mathbf{k}, \omega) + \varepsilon_r^{(i)}(\mathbf{k}, \omega) - 1} \right), \quad (3.3)$$

with superscripts (e) and (i) specifying electrons and ions. The asymmetry between the two terms is caused by lacking "unity" from the reference ions that induce the electron dress. In the denominators the full $(\mathbf{k}, \omega)$-dependent dielectric function (D.5) is recognized.

The power spectrum of density fluctuations

To obtain the power spectrum of the density fluctuations we need to find $\langle n(\mathbf{k}, \omega) n^*(\mathbf{k}', \omega') \rangle$. The random element will come from the distributions of reference electrons and ions, while the "dress" is a deterministic part following them in the present model. This latter part is simple to deal with. Here the mean square of the random part discussed in more detail [22]. The velocities $\mathbf{u}$ and particle positions $\mathbf{r}_0$ are randomly distributed, and statistically independent. The joint probability density of positions and velocities is then a product of the corresponding probability densities and the averages can be evaluated individually. The averaging is carried out first with respect to the randomly and uniformly distributed positions. By this it is implied that the probability of finding a particular $\mathbf{r}_{0j}$ in a small volume element $dx\,dy\,dz$ taken out of a large volume V is $dx\,dy\,dz/V$.

$$(2\pi)^2 \sum_{j=1}^{N} \sum_{\ell=1}^{N} \delta(\omega - \mathbf{k} \cdot \mathbf{u}_j) \delta(\omega' - \mathbf{k}' \cdot \mathbf{u}_\ell) \left\langle e^{i\mathbf{k} \cdot \mathbf{r}_{0j}} e^{-i\mathbf{k}' \cdot \mathbf{r}_{0\ell}} \right\rangle. \quad (3.4)$$

The $N(N-1)$ terms with $j \neq \ell$ disappear in the limit of large N since $\mathbf{r}_{0j}$ and $\mathbf{r}_{0\ell}$ are statistically independent for this case. The N terms that are left have $j = \ell$ giving

$$\left\langle e^{i\mathbf{k}\cdot\mathbf{r}_{0\ell}} e^{-i\mathbf{k}'\cdot\mathbf{r}_{0\ell}} \right\rangle = \frac{1}{V}(2\pi)^2 \delta(\mathbf{k} - \mathbf{k}'),$$

giving the contribution of this term to an expanded form of the power spectrum of the electron density fluctuations $\langle n(\mathbf{k}, \omega)n^*(\mathbf{k}', \omega')\rangle$ in the form

$$(2\pi)^2 \delta(\mathbf{k} - \mathbf{k}')\delta(\omega - \omega')\left\langle \sum_{j=1}^{N} \delta(\omega - \mathbf{k}\cdot\mathbf{u}_j) \right\rangle \frac{1}{V}$$

$$= n_0 (2\pi)^2 \delta(\mathbf{k} - \mathbf{k}')\delta(\omega - \omega') \iiint_{-\infty}^{\infty} \delta(\omega - \mathbf{k}\cdot\mathbf{u}) f_0(\mathbf{u})d^3u, \quad (3.5)$$

including now the averaging over also electron velocities, again having $n_0 = N/V$. The ensemble averaging is here over all particle velocities, with $f_0(\mathbf{u})$ again being the particle velocity distribution. It is normalized, $\iiint f_0(\mathbf{u})d^3u = 1$, but not yet specified. We also used (2.6). The power spectrum of the electron density fluctuations is then obtained as

$$\langle n(\mathbf{k}, \omega)n^*(\mathbf{k}, \omega)\rangle = n_0 V \left| 1 - \frac{\varepsilon_r^{(e)}(\mathbf{k}, \omega) - 1}{\varepsilon_r^{(e)}(\mathbf{k}, \omega) + \varepsilon_r^{(i)}(\mathbf{k}, \omega) - 1} \right|^2$$

$$\times \iiint_{-\infty}^{\infty} \delta\left(\omega - \mathbf{k}\cdot\mathbf{u}^{(e)}\right) f_0^{(e)}(\mathbf{u})d^3u^{(e)}$$

$$+ n_0 V \left| \frac{\varepsilon_r^{(e)}(\mathbf{k}, \omega) - 1}{\varepsilon_r^{(e)}(\mathbf{k}, \omega) + \varepsilon_r^{(i)}(\mathbf{k}, \omega) - 1} \right|^2$$

$$\times \iiint_{-\infty}^{\infty} \delta\left(\omega - \mathbf{k}\cdot\mathbf{u}^{(i)}\right) f_0^{(i)}(\mathbf{u})d^3u^{(i)}. \quad (3.6)$$

There are no "cross terms" since dressed electrons and ions were assumed to move independently. The averages of the corresponding products will then vanish. As mentioned before, there is a lack of symmetry in the electron and ion contribution due to the number 1 appearing for the electrons and not for the ions. This is explained (again) by the observation that only electrons scatter, and not ions due to their large mass. The number 1 is all that remains if you assume scattering from randomly distributed independent electrons. If all $\varepsilon_r^{(e,i)}(\mathbf{k}, \omega) \to 1$ as for vacuum, the result for scattering by non-interacting independently moving electrons is recovered. If the ions are assumed to be cold, $f_0^{(i)}(\mathbf{u}) = \delta(\mathbf{u})$, with $\varepsilon_r^{(e)} \neq 1$ we find that the last term can contribute, although details have to wait until $\varepsilon_r^{(e)}$ and $\varepsilon_r^{(i)}$ are known. At first sight this can seem counter-intuitive, but even a fixed ion will influence the electron population by its mere presence: it will be Debye-screened by an electron cloud which in turn contributes to the scattering cross section. The expression (3.6) gives the frequency resolved contribution to the radar backscatter given $\mathbf{k}$. The total backscatter is found by integrating over all ω.

Inspecting (3.6) the relative importance of the two terms on the right hand side can be distinguished. Assuming that the velocity distributions entering have the form resembling Maxwellians (not necessarily *being* one) it can be argued that for $\omega/k \gg u_{thi}$, the ion thermal velocity, $\iiint_{-\infty}^{\infty} \delta(\omega - \mathbf{k} \cdot \mathbf{u}^{(i)}) f_0^{(i)}(\mathbf{u}) d^3 u^{(i)}$ is small compared to $\iiint_{-\infty}^{\infty} \delta(\omega - \mathbf{k} \cdot \mathbf{u}^{(e)}) f_0^{(e)}(\mathbf{u}) d^3 u^{(e)}$. For small for $\omega/k \ll u_{the}$ the term $\iiint_{-\infty}^{\infty} \delta(\omega - \mathbf{k} \cdot \mathbf{u}^{(e)}) f_0^{(e)}(\mathbf{u}) d^3 u^{(e)}$ is found to be small compared to the ion contribution in that parameter range. Both terms in (3.6) contribute for all $\{\omega, \mathbf{k}\}$, but the contribution from the first electron term dominates for $\omega/k \gtrsim u_{the}$ while the last ion term dominates for $\omega/k \sim C_s$, the ion sound speed, unless strongly distorted velocity distributions are considered. The relative dielectric functions $\varepsilon_r^{(e)}(\mathbf{k}, \omega)$ and $\varepsilon_r^{(i)}(\mathbf{k}, \omega)$ can usually be simplified in these two limits. For large frequencies, in particular, it can be expected that $\varepsilon_r^{(i)}(\mathbf{k}, \omega) \sim 1$ as for vacuum due to the large ion inertia. Similarly, it can be expected that electron inertia can be ignored for low frequencies. The latter assumption will usually imply that the electron component can be assumed to be in local isothermal Boltzmann equilibrium [29]. The assumption is not implying quasi-neutrality, see Appendix C, nor is imposing restrictions on the ion velocity distribution.

It was mentioned earlier that the phase information of individual Fourier contributions is lost when obtaining the auto correlation function and the power spectrum derived from it. By the notation $\langle n(\mathbf{k}, \omega) n^*(\mathbf{k}, \omega) \rangle$ for the power spectrum, this might become even more evident.

Provided we know the respective dielectric functions $\varepsilon_r^{(e)}(\mathbf{k}, \omega)$ and $\varepsilon_r^{(i)}(\mathbf{k}, \omega)$ together with the velocity distribution functions $f_0^{(e)}(\mathbf{u})$ and $f_0^{(i)}(\mathbf{u})$, we are principle finished. It will be demonstrated that the information concerning the velocity distribution functions suffices to determine also the dielectric response functions for plasmas with large N_p. Using a Maxwellian form for f_0 we expect to recover the expression (2.8) for ions and electrons by inserting the proper thermal velocities when the collective interactions are ignored as assumed in the derivation of (2.8).

3.1.3 The Kinetic Plasma Dielectric Function

So far it was assumed the total dielectric function $\varepsilon_r(\mathbf{k}, \omega)$ is known somehow. For plasma media it can be found analytically in certain limits, in particular for plasmas with large plasma parameters N_p. The fluid model is inaccurate by omitting Landau damping. To include this you need to use the full kinetic plasma dielectric function derived from the collisionless Vlasov equation [27, 30]. The analysis is entirely similar for the electrons and ions, so a summary of only the electron calculations are presented here.

The collisionless Vlasov equation can be written in the form

$$\frac{\partial f}{\partial t} + \mathbf{u} \cdot \nabla f + \frac{\mathbf{K}}{m} \cdot \nabla_{\mathbf{u}} f = 0, \tag{3.7}$$

where $f = f(\mathbf{r}, \mathbf{u}, t)$ is the distribution of plasma particles with mass m, moving under the action of a force field $\mathbf{K} = \mathbf{K}(\mathbf{r}, t)$. By $\nabla_{\mathbf{u}}$ we understand the vector operator $\{\partial/\partial u_x, \partial/\partial u_y, \partial/\partial u_z\}$. The force $\mathbf{K}$ can be prescribed or determined self-consistently through Maxwell's equations in terms of charge and current distributions in the plasma. For electrostatic dynamics we have $\mathbf{K} = q\mathbf{E}$ with q being the charge of the species, in particular $q = -e$ for electrons. It can be argued that the term "electrostatic" is a misnomer here, after all we are discussing wave-like dynamic motions, but the electric field $\mathbf{E}(\mathbf{r}, t)$ can be determined by Poisson's equation just as in electrostatics. For electrostatic waves, the plasma currents must cancel Maxwell's displacement current in Ampere's law. Otherwise there would be a source for fluctuating magnetic fields. Sources of electrostatic waves are charge fluctuations and they have wave vectors parallel to the electric field, $\mathbf{k} \parallel \mathbf{E}$. The charge fluctuations are coupled to current fluctuations through the continuity equation. Transverse electromagnetic waves have $\mathbf{k} \perp \mathbf{E}$ with no density fluctuations associated. For magnetized plasmas in particular, waveforms exist that have both longitudinal and transverse components [30–32], the latter decreasing near resonances where $|\mathbf{k}| \to \infty$. For finite wavenumbers, the waves can be quasi-longitudinal [31].

The Vlasov equation describes a phase space continuum with $f(\mathbf{r}, \mathbf{u}, t)$ assumed to be a continuous function. In reality a distribution function describes the dynamics of many δ-functions accounting for particle positions and velocities, electrons or ions. To allow a continuum model the density of these δ-function has to be large. A formally similar problem is found when considering a drop of ink in a glass of water. By inspection from a distance the ink will form a continuum, and the identification of the individual tiny grains is possible only at a large magnification. The particle density in phase space is measured by their number in a Debye sphere having the Debye length λ_D as radius, giving large plasma parameters N_p for relevant plasmas, see Appendix C.1.5 for definitions. Ideally, the Vlasov limit assumes $N_p \to \infty$.

The last term in (3.7) is nonlinear and to make progress of substance here, it is necessary to linearize the equation. Any velocity distribution of the form $n_0 f_0(\mathbf{u})$ is a solution to the equation with the convenient normalization $\iiint f_0(\mathbf{u}) d^3 u = 1$. Introduce a small perturbation $f_1 \ll f_0$ so that $f = n_0 f_0 + f_1$. Ignoring products of small terms the linear form for the electron Vlasov equation becomes

$$\frac{\partial f_1}{\partial t} + \mathbf{u} \cdot \nabla f_1 - \frac{e n_0}{m} (\mathbf{u} \times \mathbf{B}_1 + \mathbf{E}_1) \cdot \nabla_{\mathbf{u}} f_0 = 0. \tag{3.8}$$

Assuming electrostatic waves from the outset we might set $\mathbf{B}_1 = 0$ here, but note that for an isotropic velocity reference or unperturbed distribution function $f_0 = f_0(|\mathbf{u}|)$ we have $\nabla_{\mathbf{u}} f_0(u) = \widehat{\mathbf{u}} \, df_0(u)/du$ so that $\mathbf{u} \times \mathbf{B}_1 \cdot \nabla_{\mathbf{u}} f_0 = 0$ anyhow. Without the assumption of electrostatic waves, a fluctuating magnetic field would be found in Faraday's law even in this latter case. Equation (3.8) is to be completed with an expression for the relation between $\mathbf{E}_1$ and the distribution function $f_1(\mathbf{r}, \mathbf{u}, t)$.

The Vlasov equation (3.7) is time reversible for all physically relevant force terms. (The reader is urged to check this for the subtle case including the Lorentz force.) The Landau damping is thus *not* associated with anything like classical irreversible dissipation. The literature [29, 30, 33] contains detailed discussions and explanations of this phenomenon.

3.1.4 Integration Along Unperturbed Orbits

A solution of (3.8) can be written in the form

$$f_1(\mathbf{r}, \mathbf{u}, t) = f_1(\mathbf{r} - \mathbf{u}t, \mathbf{u}, 0) + \frac{en_0}{m} \int_0^t \mathbf{E}_1 \left(\mathbf{r} - \mathbf{u}(t - t'), t' \right) \cdot \nabla_{\mathbf{u}} f_0(\mathbf{u}) dt' \quad (3.9)$$

where the initial value of $f_1(\mathbf{r}, \mathbf{u}, 0)$ appears explicitly. The proof is left to reader as an exercise. In (3.9) we recognize the *characteristics* $\boldsymbol{\xi} \equiv \mathbf{r} - \mathbf{u}t$ of the linear Vlasov equation (3.8). These are the curves followed by particles moving along unperturbed straight-line orbits, and the solution (3.9) is often termed "integration along unperturbed orbits". In can be seen as a mathematical argument for the electron orbits used in Sect. 2.1.

3.1.5 Derivation of the Kinetic Dielectric Function

Introducing the electrostatic potential by $\mathbf{E}_1 = -\nabla \phi$ and Fourier transforming with respect to temporal and spatial variables we find

$$(\omega - \mathbf{k} \cdot \mathbf{u}) f_1(\mathbf{k}, \omega, \mathbf{u}) = \frac{en_0}{m} \phi(\mathbf{k}, \omega) \, \mathbf{k} \cdot \nabla_{\mathbf{u}} f_0(\mathbf{u}) \quad (3.10)$$

The electron density is found by integration

$$n(\mathbf{k}, \omega) = \iiint_{-\infty}^{\infty} f_1(\mathbf{k}, \omega, \mathbf{u}) d^3 u = \frac{en_0}{m} \phi(\mathbf{k}, \omega) \iiint_{-\infty}^{\infty} \frac{\mathbf{k} \cdot \nabla_{\mathbf{u}} f_0(\mathbf{u})}{\omega - \mathbf{k} \cdot \mathbf{u}} d^3 u, \quad (3.11)$$

where the singularity at $\omega = ku$ is dealt with by the Landau contour of integration [27, 34], see Fig. 3.2. The singularity is found only for velocity components along $\mathbf{k}$, the two other velocity integrations are easy.

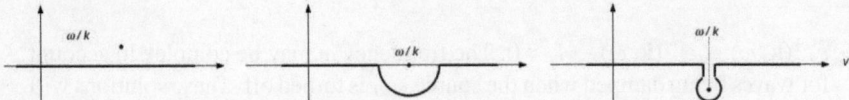

Fig. 3.2 Illustration of the Landau contour for varying ω/k in a complex velocity plane

Recall the general definition of the relative dielectric function of a dispersive medium as given by

$$\varepsilon_r(k, \omega) = 1 + \frac{en(k, \omega)}{\varepsilon_0 k^2 \phi(k, \omega)},$$

again in a Fourier transformed version, see (D.3).

The relative complex kinetic plasma dielectric function is obtained as

$$\varepsilon_r(k, \omega) = 1 - \frac{\omega_{pe}^2}{k^2} \oint_{-\infty}^{\infty} \frac{f_0'(u_x)}{u_x - \omega/k} du_x. \tag{3.12}$$

For real ω and real k we have in particular

$$\varepsilon_r(k, \omega) = 1 - \frac{\omega_{pe}^2}{k^2} P \int_{-\infty}^{\infty} \frac{f_0'(u_x)}{u_x - \omega/k} du_x - i\pi \frac{\omega_{pe}^2}{k^2} f_0'(\omega/k). \tag{3.13}$$

The dielectric function is complex with an imaginary part, which for the case $f_0'(\omega/k) < 0$ implies some unspecified dissipation: here it is Landau damping. Physically, this implies that in order to maintain the real frequency ω, it is necessary to supply energy to the system. For small k the damping will usually be small since $\omega \approx \omega_{pe}$ so that $f_0'(\omega/k \to \infty) \to 0$ when $k \to 0$.

The kinetic plasma dispersion function

The one dimensional form (3.13) is seen most often. It can be presented in a fully three dimensional version as

$$\varepsilon_r(\mathbf{k}, \omega) = 1 + \frac{\omega_{pe}^2}{k^2} P \iiint_{-\infty}^{\infty} \frac{\mathbf{k} \cdot \nabla_{\mathbf{u}} f_0(\mathbf{u})}{\omega - \mathbf{k} \cdot \mathbf{u}} d^3 u$$

$$- i\pi \frac{\omega_{pe}^2}{k^2} \iiint_{-\infty}^{\infty} \delta(\omega - \mathbf{k} \cdot \mathbf{u}) \mathbf{k} \cdot \nabla_{\mathbf{u}} f_0(\mathbf{u}) d^3 u. \tag{3.14}$$

The dispersion relation for the electrostatic plasma waves is generally found by $\varepsilon_r(\mathbf{k}, \omega) = 0$. Only in this case do we find finite electric fields without external charges to maintain them, as seen from Poisson's equation in the form $i\mathbf{k} \cdot \varepsilon_0 \varepsilon_r(\mathbf{k}, \omega) \mathbf{E}(\mathbf{k}, \omega) = \zeta_{ext}(\mathbf{k}, \omega)$. Significant contributions to the fluctuating density power spectrum (3.6) thus come from $(\mathbf{k}, \omega)$-combinations where

$\varepsilon_r^{(e)}(\mathbf{k}, \omega) + \varepsilon_r^{(i)}(\mathbf{k}, \omega) - 1 \approx 0$. The frequency ω may be complex to account for waves being damped when the source ζ_{ext} is turned off. These solutions will correspond to high frequency Langmuir waves and low frequency ion sound waves for the present unmagnetized conditions. The power spectrum will be considerably more complicated when an external magnetic field is imposed [35], but the unmagnetized results will apply for the direction $\parallel \mathbf{B}$.

3.1.6   The Z-Function

For arbitrary velocity distribution functions, the integral in the dielectric function has to be evaluated numerically, in general. For the special case where the velocity distribution is a Maxwellian, a tabulated function can be used [36], the so called Z-function and its derivative Z', see also Appendix E.1. We proceed to express

$$\varepsilon_r(k, \omega) = 1 - \left(\frac{\omega_p}{k}\right)^2 \fint_{-\infty}^{\infty} \frac{f_0'(u)}{u - \omega/k} \, du, \tag{3.15}$$

taking the special case where the velocity distribution is a Maxwellian

$$f_0(u) = \sqrt{\frac{m}{2\pi T}} \, e^{-\frac{1}{2}mu^2/T}. \tag{3.16}$$

At this place, m can be the mass of any of the plasma constituents, and similarly for the temperature T. The plasma frequency ω_p in (3.15) is obtained for the appropriate species, see Appendix C.1. The symbol $\fint$ for the Landau contour of integration is used again. Insertion and use of the definition of the electron Debye length λ_{De} given in Appendix C.1 gives

$$\varepsilon_r(k, \omega) = 1 + \frac{1}{k^2 \lambda_{De}^2 \sqrt{\pi}} \fint_{-\infty}^{\infty} \frac{x e^{-x^2}}{x - z} \, dx$$

$$= 1 + \frac{1}{k^2 \lambda_{De}^2} + \frac{z}{k^2 \lambda_{De}^2 \sqrt{\pi}} \fint_{-\infty}^{\infty} \frac{e^{-x^2}}{x - z} \, dx, \tag{3.17}$$

with the normalized variable

$$z \equiv \frac{\omega}{k} \sqrt{\frac{m}{2T}},$$

using $\int_{-\infty}^{\infty} \exp(-x^2) dx = \sqrt{\pi}$.

The special or limiting case of $\omega/k = 0$ can readily be found as

$$\varepsilon_r(k, 0) = 1 + \frac{1}{k^2 \lambda_{De}^2} \tag{3.18}$$

The inverse Fourier transform [21] for the potential found by Poisson's equation is in this case

$$InvFT\left\{\frac{q}{k^2 \varepsilon(k, 0)}\right\} = q \exp\left(-\frac{r}{\lambda_{De}}\right) \quad \text{for} \quad r > 0,$$

demonstrating that (3.18) corresponds to the electron Debye screening, here in 1 spatial dimension, as different from the 3 dimensional version found in (C.6).

Because of the frequent use of (3.17) in plasma physics, it is found useful to introduce a function, often called the "Z-function", defined as

$$Z(z) \equiv \frac{1}{\sqrt{\pi}} \fint_{-\infty}^{\infty} \frac{e^{-x^2}}{x - z} dx. \tag{3.19}$$

It is found that $Z(z)$ is the Hilbert transform [21, 37] of $\exp(-x^2)$. More generally, the plasma dispersion relation involves the Hilbert transform of the plasma velocity distribution function. It is found by construction that the Kramers-Kronig relations [21, 32] for absorption and refraction are automatically satisfied by this result for collisionless plasmas. It would have been unfortunate if it was not so, but it is nonetheless interesting that the the result applies for Landau damping which has its origin in time-reversible processes [30, 34, 38].

Also the derivative $Z'(z)$ of the $Z(z)$-function will be needed.

$$Z'(z) = \frac{1}{\sqrt{\pi}} \fint_{-\infty}^{\infty} \frac{e^{-x^2}}{(x - z)^2} dx = -\frac{1}{\sqrt{\pi}} \fint_{-\infty}^{\infty} e^{-x^2} \frac{d}{dx} \frac{1}{x - z} dx$$

$$= -\frac{2}{\sqrt{\pi}} \fint_{-\infty}^{\infty} \frac{xe^{-x^2}}{x - z} dx \tag{3.20}$$

after a partial integration. To demonstrate the relation to the plasma dielectric function we define

$$I(z) \equiv \frac{1}{\sqrt{\pi}} \fint_{-\infty}^{\infty} \frac{xe^{-x^2}}{x - z} dx = 1 + \frac{z}{\sqrt{\pi}} \fint_{-\infty}^{\infty} \frac{e^{-x^2}}{x - z} dx. \tag{3.21}$$

which is a normalized version of the second part of $\varepsilon_r(k, \omega)$ in (3.17). The integral in the last term in (3.21) is the Z-function

$$Z'(z) = -2(1 + zZ(z)), \tag{3.22}$$

see also Appendix E.1.

The longitudinal plasma dispersion function (3.17) for a plasma in thermal equilibrium at a temperature T can now be expressed as

$$\varepsilon_r(k, \omega) = 1 - \frac{1}{2(k\lambda_{De})^2} Z'\left(\frac{\omega}{k}\sqrt{\frac{m}{2T}}\right). \tag{3.23}$$

The equation $\varepsilon_r(k, \omega) = 0$ with (3.23) has complex solutions $\omega = \mathcal{R}e\{\omega\} + i\,\mathcal{I}m\{\omega\}$ for a given k, see illustrations in Fig. 3.3. Note the increase in damping rate, i.e., $|\mathcal{I}m\{\omega\}|$, for increasing k. Also note that $\mathcal{R}e\{\omega\} \approx 1$, i.e., ω is close to the plasma frequency ω_{pe} for small k. In case one of the zeroes is much closer to the $\mathcal{I}m\{\omega\} = 0$ line than any of the other ones, then the time asymptotic wave-field will be dominated by the contributions from that solution. This will be the case for Langmuir waves in the limit of small wavenumbers, but not necessarily so for ion acoustic waves where several zeroes can have comparable imaginary parts [39].

The derivation of the contribution from Maxwellian ions with temperature T_i to the dielectric function follows the same lines as the analysis for electrons.

Generally, inclusion of both the electron and the ion dynamics by (D.5) gives

$$\varepsilon_r(k, \omega) = 1 - \left(\frac{\omega_{pe}}{k}\right)^2 \fint_{-\infty}^{\infty} \frac{f'_{0e}(u)}{u - \omega/k}\,du - \left(\frac{\Omega_{pi}}{k}\right)^2 \fint_{-\infty}^{\infty} \frac{f'_{0i}(u)}{u - \omega/k}\,du. \tag{3.24}$$

For Maxwellian plasmas with no relative drift velocities the result is

$$\varepsilon_r(k, \omega) = 1 - \frac{1}{2(k\lambda_{De})^2} Z'\left(\frac{\omega}{k}\sqrt{\frac{m}{2T_e}}\right) - \frac{1}{2(k\lambda_{Di})^2} Z'\left(\frac{\omega}{k}\sqrt{\frac{M}{2T_i}}\right),$$

recalling the definition of the variable z.

A simplified limit can be found for $(\omega/k)\sqrt{m/2T_e} \ll 1$, while $(\omega/k)\sqrt{M/2T_i}$ is unspecified but finite. The electron contribution to the dielectric function here becomes $\varepsilon_r^{(e)} = 1 + (k\lambda_{Di})^{-2}$. Physically, this corresponds to the assumption that the electrons at all times can maintain a local isothermal Boltzmann equilibrium at a temperature T_e. For $\varepsilon_r^{(e)}(\mathbf{k}, \omega) + \varepsilon_r^{(i)}(\mathbf{k}, \omega) - 1 \equiv \varepsilon_r(\mathbf{k}, \omega)$ we then have

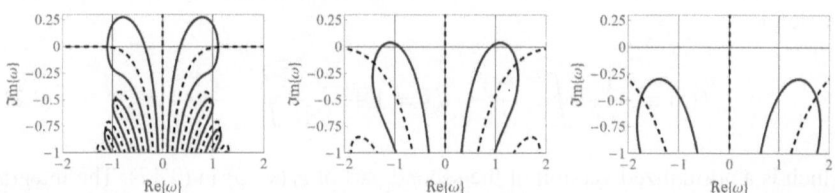

Fig. 3.3 Solutions of the implicitly given normalized dispersion relation $\varepsilon_r(k, \omega) = 0$ with (3.23), here for $k = 0.25, 0.50$ and 0.75. Full lines gives the zero level for the real part and dashed lines for the imaginary part. The crossing of two such lines gives a solution for the full dispersion relation. There are many solutions. The least damped one is most relevant for describing the propagating wave observations

$$\varepsilon_r(\mathbf{k}, \omega) \approx 1 + \frac{1}{(k\lambda_{De})^2} - \frac{1}{2(k\lambda_{Di})^2} Z'\left(\frac{\omega}{k}\sqrt{\frac{M}{2T_i}}\right). \tag{3.25}$$

By the second term, this model retains the electron Debye shielding of a fixed charge. In the quasi neutral limit, see Appendix C, the number 1 can be ignored as compared to $(k\lambda_D)^{-2}$ on the right hand side of (3.25) to give

$$\varepsilon_r(\mathbf{k}, \omega) \approx \frac{1}{(k\lambda_{De})^2}\left(1 - \frac{T_e}{2T_i} Z'\left(\frac{\omega}{k}\sqrt{\frac{M}{2T_i}}\right)\right). \tag{3.26}$$

The second term in (3.6) can be simplified in this limit to give

$$\left|\frac{\varepsilon_r^{(e)}(\mathbf{k}, \omega) - 1}{\varepsilon_r^{(e)}(\mathbf{k}, \omega) + \varepsilon_r^{(i)}(\mathbf{k}, \omega) - 1}\right|^2 \rightarrow \frac{1}{\left|1 - \frac{T_e}{2T_i} Z'\left(\frac{\omega}{k}\sqrt{\frac{M}{2T_i}}\right)\right|^2}.$$

Use of the expansion for small arguments of the Z-function, see Appendix E, gives the approximation

$$\frac{1}{\left|1 - \frac{T_e}{2T_i} Z'\left(\frac{\omega}{k}\sqrt{\frac{M}{2T_i}}\right)\right|^2} \approx \frac{1}{\left|1 + \frac{T_e}{T_i}\left(1 - \left(\frac{\omega}{k}\right)^2 \frac{M}{T_i} + i\frac{\omega}{k}\sqrt{\frac{\pi M}{2T_i}}\right)\right|^2}, \tag{3.27}$$

to a lowest approximation in ω. Similar approximations can be found for the full $\varepsilon_r(\mathbf{k}, \omega)$ for a thermal plasma with no net currents, allowing for $T_e/T_i \neq 1$.

3.1.7 Illustrative Examples

The model (3.25) is restrictive, but applies for plasmas near thermal equilibrium, allowing for $T_e \neq T_i$. It will fail, for instance, to account for current carrying plasmas with relative drifts between the electron and ion components.

As mentioned, the Z-function is tabulated [36] and can also be programmed by use of the complex error function, see Appendix E.1. Simple approximate expressions are found as well [30]. The desired Z'-function is obtained as shown before, see (3.22). For Maxwellian distributions, or sums of such, the relative dielectric function $\varepsilon_r(k, \omega)$ can be calculated for collisionless plasmas to give the power spectrum for the electron density fluctuations. The dominant contribution to the density power spectrum near the electron plasma frequency is shown in Fig. 3.4. The $k\lambda_D$-variation can be interpreted in different ways: for fixed plasma conditions (i.e., fixed λ_D) we have the variation due to changes in the radar wavelengths. Alternatively we can let k be fixed, and then the figure shows the variation of the density power spectrum due to changes in plasma parameters, the electron temperature in particular, as they are

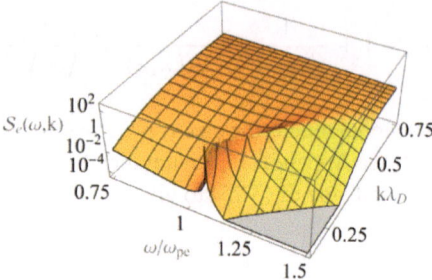

Fig. 3.4 Illustration of a power spectrum for density fluctuations shown on normalized logarithmic scales. The figure shows only the dominant contribution for the Langmuir branch in (3.6). The Langmuir peak becomes high and narrow for $k\lambda_D$ and is difficult to represent graphically

manifested through changes in λ_D. Changes in plasma density give corresponding variations in λ_D, but in that case the normalization on the frequency axis changes also. Note the steady increase of $S_e(\omega, k)$ for increasing $k\lambda_D$ when $\omega < \omega_{pe}$ is fixed.

With the approximations mentioned previously, see Sect. 3.1.2, we find here that the result (3.2) gives an adequate approximation. The dispersive Langmuir dispersion relation can be identified and also the electron Landau damping that quenches the waves when $k\lambda_D \gtrsim 0.5$. Organized motions can not be maintained for Langmuir-wavelengths comparable to or smaller than the Debye length. In this limit the plasma particles can be assumed to move independently. More detailed arguments are given in Appendix C.1. For radar wavelengths comparable to or smaller than λ_D the incoherent scattering will actually be from independent particles and the results from Chap. 2 will apply as they are, as for "undressed" electrons.

For the frequency region covering the ion sound regime, the use of (3.25) gives the results shown in Fig. 3.5 for two electron-ion temperature ratios. The sound wave dispersion relation can be identified for the case with $T_e/T_i = 2$ and also the heavy ion Landau damping that sets in as soon as $k\lambda_D \gtrsim 1$. Also ion sound waves become

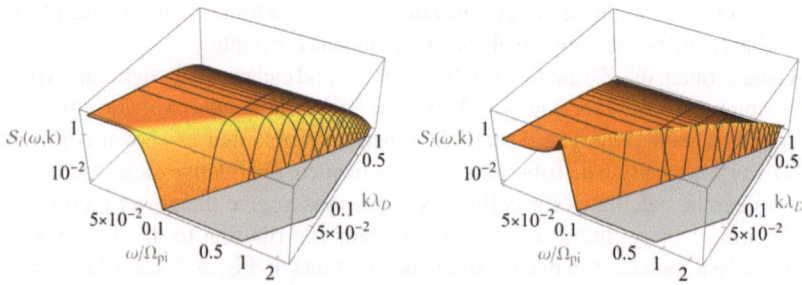

Fig. 3.5 Illustration of a power spectrum for density fluctuations shown on normalized logarithmic scales for the dominant contribution to (3.6) at the ion sound branch for temperature ratios $T_e/T_i = 1$ and 5. Due to the heavy ion Landau damping, the sound waves are almost indistinguishable for the smallest temperature ratio

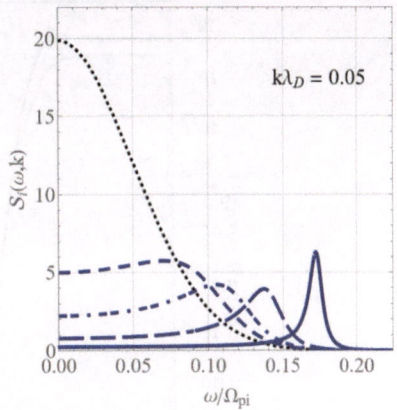

Fig. 3.6 Illustration of low-frequency spectral changes for changing electron-ion temperature variations, here shown for $T_e/T_i = 0, 1, 2, 4, 8$, with dotted, dashed, double-dashed, long-dashed and full lines, respectively

dispersive for increasing $k\lambda_D$ but this feature can be noted only for large temperature rations where the ion Landau damping is reduced. For $\omega < kC_s$ a slight decrease of $S_i(\omega, k)$ can be found when $k\lambda_D$ increases.

The consequences of changes in electron-ion temperature ratios, T_e/T_i, are illustrated in Fig. 3.6 for a fixed wavenumber, $k\lambda_D = 0.05$. A result for $T_e = 0$ is inserted for reference with a dotted line, although it will hardly be found in nature. For $T_e = T_i$ the ion Landau damping is strong and the ion sound wave will be heavily damped. Due to the increase in sound speed the frequency of the peak corresponding to the ion sound wave is increasing with T_e/T_i and the peak gets narrower corresponding to a weaker ion Landau damping. An outline and explanation of the physical mechanism behind the sound speed variation with varying electron pressure is found in the literature [29, 30].

A sample of results for a full density power spectrum, including both the electron and ion lines, is shown in Fig. 3.7 in a double logarithmic presentation with $1/k\lambda_{De}$ as a parameter. A pronounced peak around the electron plasma frequency can be noted when $1/k\lambda_{De}$ is large, giving weak Landau damping. A small "hump" is found for small frequencies, this is the remnants of the ion line. The electron-ion temperature ratio is not large enough here to give a distinct peak for the sound speed.

The effect of a relative electron-ion drift is illustrated in Fig. 3.8 for several values of the electron drift velocity with parameters $T_e/T_i = 2$, $k\lambda_D = 0.05$ and an ion-electron mass ratio of $M/m = 2 \times 10^3$. Due to the relative drift, an asymmetry develops between the positive and negative Doppler shifts. The asymmetry increases steadily for increasing velocities U. The difference is due to the modification of the electron Landau damping on the ion sound waves. The simplest explanation of the effect can be found by noting the change in the slope of the electron velocity distribution at velocities $u \approx C_s$. The final result is due to a "competition" between the ion and electron Landau dampings.

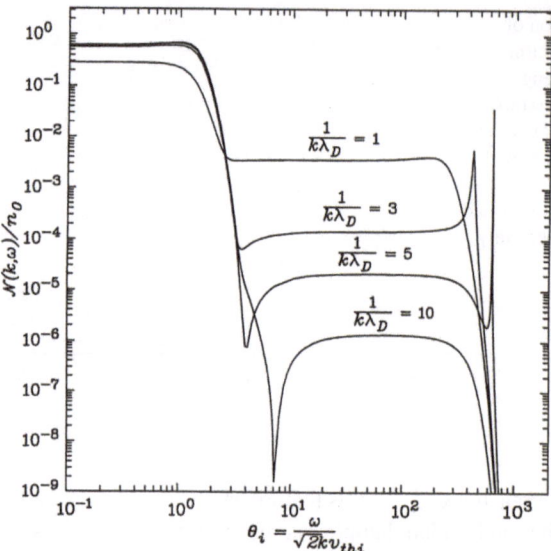

Fig. 3.7 Illustration of a power spectrum for density fluctuations shown on a double logarithmic scale as a function of normalized frequency for different values of $1/k\lambda_D$. The electron-ion mass ratio is chosen so that $\sqrt{M/m} = 172$ and $T_e/T_i = 1$ giving a heavy ion Landau damping. The narrow peak at large frequencies comes from the Langmuir waves [40], the broad contribution with a small hump originates from ion sound waves. The figure is taken from lecture notes by Prof. Tor Hagfors at the EISCAT Summer School in 1996 [11], and reproduced here by the kind permission of the EISCAT Scientific Association. All rights reserved. Figures with related information can be found also in the literature [8, 20, 22]

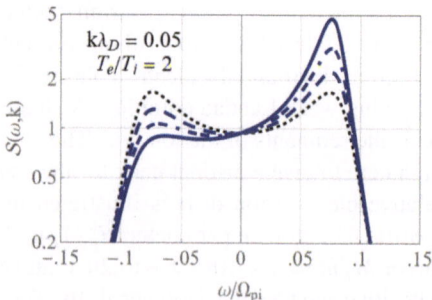

Fig. 3.8 Illustration of the consequences of an electron drift with normalized velocity U with respect to ions at rest. The temperature ratio is $T_e/T_i = 2$ and $k\lambda_D = 0.05$. The black dotted line gives the reference for $U = 0$, while the other curves are for $U = 0.1, 0.2$ and 0.3 (with dashed, dot-dashed and full lines, respectively) in units of the electron thermal velocity

3.1.8 Kappa Distributions

The foregoing summary assumed a single Maxwellian velocity distribution for the plasma particles. Combinations of several such distributions is straight forward. The various components can have different temperatures, densities, relative drifts, etc., and a variety of cases can be approximated this way. Experience has shown that photo-electrons (or primary electrons in laboratory discharge plasmas) are described better by the so-called kappa-distributions [41–45]. They appear like thermal distributions with an enhanced density "tail" of hot plasma, and are representative for some conditions found in nature, the ionosphere in particular. The analytical form is

$$F_\kappa(u_\parallel, u_\perp) = \frac{n}{\pi^{3/2}\Theta_\parallel\Theta_\perp^2}\frac{\Gamma(\kappa+1)}{\kappa^{3/2}\Gamma(\kappa-1/2)}\left(1+\frac{u_\parallel^2}{\kappa\Theta_\parallel^2}+\frac{u_\perp^2}{\kappa\Theta_\perp^2}\right)^{-(\kappa+1)} \tag{3.28}$$

where $\Theta_\parallel = [(2\kappa-3)/\kappa]^{1/2}(T_\parallel/m)^{1/2}$ and $\Theta_\perp = [(2\kappa-3)/\kappa]^{1/2}(T_\perp/m)^{1/2}$, with $\Theta_\parallel^2$ and $\Theta_\perp^2$ being effective temperatures [46] when $\kappa > 3/2$ and n is the local density of the relevant species. Temperatures are also here given with the Boltzmann constant included to avoid confusion with κ. See Fig. 3.9 for illustrations on semi-logarithmic axes. The analysis giving the Z-function can be generalized for such distributions. Mixed kappa-Maxwellian distributions in particular have been suggested [47]. An equivalent for the Z-function for κ-distributions can be expressed in terms of Gauss' Hypergeometric function [47]. It can be demonstrated [42] that the kappa-distribution approaches a Maxwellian in the limit $\kappa \to \infty$.

3.1.9 Visualizing the Particle Dress

The dress of a charged particle, stationary or moving is so far presented on by its Fourier transform, i.e., in $(\omega, \mathbf{k})$ space. For a physical understanding of the phenomenon it would be an advantage to have a presentation in "real physical space",

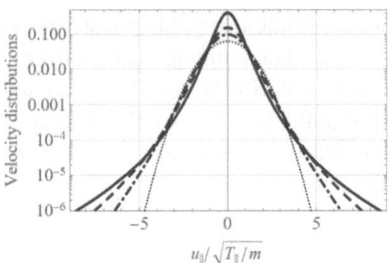

Fig. 3.9 Illustration of κ-velocity distributions (3.28) for $\kappa = 2$, 3 and 5 given with full, dashed and dash-dotted lines, respectively. The figure shows a "cut" in the velocity distributions for $u_\perp = 0$ in (3.28). A thin dotted line shows the corresponding Maxwellain distribution for the same temperature. The figure assumes $T_\parallel = T_\perp$

the configuration space, as the one shown in Fig. 3.1. It turns out that the transformation from one space to the other is complicated and lengthy, so it is placed in an Appendix F.1 in order not to disturb the presentation. For instance, the radiation pattern associated with electron plasma waves is illustrated in Fig. F.1 as obtained by a fluid model where Landau damping is ignored. This is permissible for a relevant parameter range. For parameters usually found in the ionosphere it is not permissible to ignore ion Landau damping, so that analysis requires a kinetic treatment.

3.2 Magnetized Plasmas

Imposing a homogeneous magnetic field $\mathbf{B}_0$ on the plasma gives rise to a number of complications. As already mentioned, the polarization of the electric field of the radar is no longer trivial since we now have a preferred direction given by the magnetic field. If adding two linearly polarized waves to obtain circular polarization, we find that the direction of rotation for circularly (or elliptically) polarized fields becomes important; a reference circulation is given by the gyro-rotation of the electrons. To express $\mathbf{r} = \mathbf{r}_0 + \mathbf{u}t$ it is necessary to account for the helical particle trajectories shown in Fig. 2.2. Also the dielectric functions of the plasma are modified substantially by a magnetic field.

The basic equation is again the Vlasov equation

$$\frac{\partial f}{\partial t} + \mathbf{u} \cdot \nabla f - \frac{en_0}{m} (\mathbf{u} \times \mathbf{B} + \mathbf{E}) \cdot \nabla_{\mathbf{u}} f = 0. \qquad (3.29)$$

An equilibrium uniform solution [33] to this equation has to fulfill $\mathbf{u} \times \mathbf{B} \cdot \nabla_{\mathbf{u}} f_0(\mathbf{u}) = 0$ giving the form $f_0 = f_0(u_x^2 + u_y^2, u_z)$ with $\hat{\mathbf{z}} \parallel \mathbf{B}_0$ being the unperturbed reference magnetic field, assumed homogeneous.

When written for electrons in a linearized form (3.29) becomes

$$\frac{\partial f_1}{\partial t} + \mathbf{u} \cdot \nabla f_1 - \frac{en_0}{m} \mathbf{u} \times \mathbf{B}_0 \cdot \nabla_{\mathbf{u}} f_1 + \frac{en_0}{m} (\mathbf{u} \times \mathbf{B}_1 + \mathbf{E}_1) \cdot \nabla_{\mathbf{u}} f_0 = 0. \quad (3.30)$$

The Vlasov equation describes time-reversible physics also when an external magnetic field is included. The Landau damping derived from it has a different nature as compared with the collisional effects usually associated with dissipation. The interested reader is referred to the literature on the subject [29, 30, 34, 38].

For a kinetic plasma model it can be shown [22] that with magnetized orbits in (3.6) we have to make the replacement

$$\iiint_{-\infty}^{\infty} f_0(\mathbf{u})\delta(\omega - \mathbf{k} \cdot \mathbf{u})d^3u \;\rightarrow$$

$$\iiint_{-\infty}^{\infty} \sum_j J_j^2 \left(\frac{k_\perp u_\perp}{\Omega_c} \right) f_0(\mathbf{u})\delta(\omega - j\Omega_c - k_\parallel u_\parallel)d^3u, \qquad (3.31)$$

in a self evident notation, with Ω_c being the cyclotron frequency for the species in question, while J_j is the Bessel function of order j. Note that the summation includes all cyclotron harmonics $j\Omega_c$. Fluid models will only recognize the fundamental frequencies.

The kinetic dielectric response function for a magnetized plasma [12, 27, 34, 48, 49] is given as

$$\varepsilon_r(\mathbf{k}, \omega) = 1 - \sum_\sigma \frac{\omega_{p\sigma}^2}{k^2} \sum_{n=-\infty}^{\infty} \iiint_{-\infty}^{\infty} \left(\frac{n\Omega_\sigma}{u_\perp} \frac{\partial f_{0\sigma}(u_\perp, u_\parallel)}{\partial u_\perp} + k_\parallel \frac{\partial f_{0\sigma}(u_\perp, u_\parallel)}{\partial u_\parallel} \right)$$

$$\times \frac{J_n^2\left(\frac{k_\perp u_\perp}{\Omega_\sigma}\right)}{n\Omega_\sigma + k_\parallel u_\parallel - \omega} d^3 u. \tag{3.32}$$

The summation $\sum_\sigma$ runs over the two species specified by the subscript σ, $f_{0\sigma}(u_\perp, u_\parallel)$ are the normalized reference velocity distribution functions, and Ω_σ, $\omega_{p\sigma}$ the corresponding cyclotron and plasma frequencies. The relation of (3.32) to the full dielectric tensor in collisionless plasmas can be found in the literature [27]. Electrostatic wave modes are also here identified by $\varepsilon_r(\mathbf{k}, \omega) = 0$. Inserting (3.32) into (3.6) we find a density power spectrum accounting for the effects of a homogeneous magnetic field. The assumption of spatial homogeneity is a minor restriction, for most relevant cases it is locally fulfilled within the radar scattering volume. The assumption of isotropy of the unperturbed velocity distributions with respect to the velocity component perpendicular to **B** is natural for collisionless plasmas, but can be violated under non-equilibrium conditions in collisional plasma as found in the Earth's ionosphere [50, 51].

For propagation along the magnetic field, $k_\perp = 0$, the Langmuir waves and ion sound waves are recovered. In addition perpendicular, or quasi-perpendicular, hybrid waves (upper- and lower-hybrid) will be found [29]. One important consequence of a kinetic model for magnetized plasmas as compared to a fluid version is that together with the hybrid wave mode all higher order cyclotron harmonics can be excited, i.e., Bernstein modes [27, 29]. This feature is markedly different from the basic fluid result which only recognizes the lower and upper hybrid frequencies and the ion and electron cyclotron frequencies, not their harmonics. The Appleton-Hartree dispersion diagram [30, 32] gives an adequate overview of the fluid results.

One curious feature of (3.32) is that waves propagating exactly perpendicular to $\mathbf{B}_0$, i.e., with $k_\parallel = 0$, are undamped. A paradox can be formed: start with a selected wavemode having $k_\parallel = 0$ and then let $B_0 \to 0$. The wavemode has to be undamped for any $B_0 > 0$, but should be Landau damped for the limit $B_0 = 0$! It is left to the reader to consult the literature [48] for the resolution of this paradox.

3.3   The Role of Zeroes of the Relative Dielectric Function

By inspection of (3.6) it is possible to form a qualified guess for the shape of the power spectrum of the fluctuating plasma density. Take first Poisson's equation in the form found in Sect. C.1.6, i.e., $\nabla \cdot \mathbf{D} = \nabla \cdot \varepsilon_0 \varepsilon_r \mathbf{E} = \rho_{ext}$. According to this expression, any electric field disappears as soon as ρ_{ext} is "switched off", unless $\varepsilon_r = 0$. For simple media with constant dielectric constants this results is of course not meaningful, but it will be so for the more general case with $(\omega, \mathbf{k})$-depending dielectrics as found for Volterra type relations (D.2). In this case electrostatic waves can be found, so named because they have fluctuating electric fields without any corresponding magnetic field variations [30]. The condition for their existence is $\varepsilon_r(\mathbf{k}, \omega) = 0$, which can be interpreted as an implicitly given dispersion relation $\omega = \omega(\mathbf{k})$. For a simple fluid model, the corresponding waves will propagate without damping, but in a fully kinetic model of collisionless plasma waves they will be subject to the linear Landau damping. The zeroes of the dielectric function can for a real wave vector $\mathbf{k}$ be interpreted as having a complex frequency, $\omega = \omega_r(\mathbf{k}) + i\omega_i(\mathbf{k})$, This would correspond to an initial value problem with the waves excited at some time $t = 0$. Alternatively, for a boundary value problem we can have a real frequency with a complex wave vector, $\mathbf{k} = \mathbf{k}_r(\omega) + i\mathbf{k}_i(\omega)$, where $\mathbf{k}_i$ accounts for a spatial damping. Physically, this means that we somewhere in space need an exciter corresponding to $\rho_{ext}(\mathbf{r}, t) \neq 0$ and the waves propagate away from this region to be damped with increasing distance. One such source of energy input can be the moving reference particle creating the "dress".

To interpret the back-scattered radar signal, it is necessary to search for zeroes of the full dielectric function in the denominators of (3.6): local peaks in the density power spectrum will be found for the corresponding frequency-wave vector combinations. These will give the dominant contribution to the radar back-scatter. For an unmagnetized plasma, the only candidates are the zeroes corresponding to Langmuir and ion sound waves. For a magnetized plasma you will find a surprising diversity of possible wave forms [27] as identified by the zeroes of (3.32). The zeroes for the kinetic dielectric functions are in the complex plane: the closer they are to the real axis (i.e., a weaker damping), the stronger is their signature in the density fluctuation power spectrum. The radar beam diagnoses by a real wave vector and returns a real frequency.

3.3.1   Approximate Results for Weak Dampings

The dielectric losses will give rise to damping of waves. This damping gives rise to a line-broadening of the frequency of the plasma waves, except in the limit $k \rightarrow 0$ where $\omega = \omega_{pe}$ for plasma media. The damping rate is determined by the real as well as the imaginary parts of $\varepsilon(\omega, \mathbf{k})$. Here only electrostatic waves will be considered for simplicity. Assume as before that the plasma (or any other medium for that matter) supports weakly damped electrostatic waves with a dispersion relation $\omega(\mathbf{k}) = \omega_1(\mathbf{k}) + i\omega_2(\mathbf{k})$ with $\omega_1(\mathbf{k}) \gg \omega_2(\mathbf{k})$. (Of course also strong dampings can be found, but this limit is not considered here.) As written before, these waves have to be associated with a zero of the dielectric function $\varepsilon(\omega(\mathbf{k}), \mathbf{k})$. For weakly damped waves we look for zeroes close to the real ω-axis, and approximate the relative dielectric function as

$$\varepsilon_r(\omega, \mathbf{k}) \equiv \varepsilon_1(\omega_1 + i\omega_2, \mathbf{k}) + i\varepsilon_2(\omega_1 + i\omega_2, \mathbf{k})$$

$$\approx \varepsilon_1(\omega_1, \mathbf{k}) + i\varepsilon_2(\omega_1, \mathbf{k}) - \omega_2\frac{\partial\varepsilon_2}{\partial\omega_1} + i\omega_2\frac{\partial\varepsilon_1}{\partial\omega_1}. \tag{3.33}$$

By the weak damping it was implicitly assumed that $\varepsilon_2 \ll \varepsilon_1$. The third term is taken to be small, because both ω_2 and ε_2 were assumed small, while ε_1 can be anything.

The condition for the existence of an electrostatic wave $\varepsilon_r(\omega, \mathbf{k}) = 0$ then implies $\varepsilon_1(\omega_1, \mathbf{k}) \approx 0$, which determines the real part of the dispersion relation for the waves, $\omega_1 = \omega_1(\mathbf{k})$. For electron plasma waves this dispersion relation will be $\omega^2 = \omega_{pe}^2 + Cu_{th}^2 k^2$, for ion sound waves $\omega^2 \approx C_s^2 k^2$ in terms of the electron plasma frequency ω_{pe}, electron thermal velocity u_{th}, the ion sound speed C_s and a constant C. The precise expression for C depends on the model used, fluid and kinetic analyses giving slightly different results [33, 52]. See also Appendix F.

Taking the imaginary part of (3.33) gives

$$\omega_2(\mathbf{k}) = -\varepsilon_2(\omega_1(\mathbf{k}), \mathbf{k}) \left/ \frac{\partial\varepsilon_1(\omega, \mathbf{k})}{\partial\omega}\right|_{\omega=\omega_1(\mathbf{k})}. \tag{3.34}$$

The numerator is given by the dielectric losses, while the denominator is determined by the same term which enters the expression for the energy density $\mathcal{W}$ of electrostatic waves [30, 53]. Here we have

$$\mathcal{W} = \frac{1}{2}\varepsilon_0|E|^2 \frac{\partial\big(\omega\varepsilon_1(\omega, \mathbf{k})\big)}{\partial\omega}\bigg|_{\omega=\omega_1} \approx \frac{1}{2}\varepsilon_0|E|^2 \omega \frac{\partial\varepsilon_1(\omega, \mathbf{k})}{\partial\omega}\bigg|_{\omega=\omega_1}$$

where ω_1 is again the solution to $\varepsilon_1(\omega, \mathbf{k}) = 0$. The vacuum energy density is recovered here as a special case.

The approximation (3.35) will not work for strongly damped cases. Such an example is the ion Landau damping for ion sound waves when the electron-ion temperature ratio is moderate.

The foregoing discussions were implicitly obtained for *temporally* damped waves with an initial condition where a plane wave was released and its damping subsequently followed in time. Physically, the alternative situation is just as interesting; a harmonic wave is excited at a certain spatial position by an antenna, and the damping of the wave is followed as it propagates away from the exciter. This problem can be analyzed in much the same way as the previous one. Assume now that ω is real, while $\mathbf{k} = \mathbf{k}_1 + i\mathbf{k}_2$ with $k_2 \ll k_1$ where k_2 accounts for the spatial damping $e^{-k_2 x}$ along the x-direction. We then have the expansion

$$\varepsilon_r(\omega, \mathbf{k}) \approx \varepsilon_1(\omega, \mathbf{k}_1) + i\varepsilon_2(\omega, \mathbf{k}_1) - \mathbf{k}_2 \cdot \nabla_{\mathbf{k}}\varepsilon_2(\omega, \mathbf{k}) + i\mathbf{k}_2 \cdot \nabla_{\mathbf{k}}\varepsilon_1(\omega, \mathbf{k}). \quad (3.35)$$

As before we obtain $\varepsilon_1(\omega, \mathbf{k}_1) = 0$ determining $\omega = \omega(\mathbf{k}_1)$, which is essentially the same result as before, but now the spatial damping is determined through

$$\mathbf{k}_2 \cdot \nabla_{\mathbf{k}}\varepsilon_1(\omega, \mathbf{k}_1) = -\varepsilon_2(\omega(\mathbf{k}_1), \mathbf{k}_1), \quad (3.36)$$

where in general $\nabla_{\mathbf{k}}\varepsilon_1$ need not be parallel to $\mathbf{k}_2$. In particular the relation between the temporal and the spatial dampings is found by combining (3.34) and (3.36) to give

$$\omega_2 = -\mathbf{k}_2 \cdot \nabla_{\mathbf{k}}\omega_1, \quad (3.37)$$

using

$$\mathbf{u}_g \equiv \nabla_{\mathbf{k}}\omega_1 = -\nabla_{\mathbf{k}}\varepsilon_1(\omega_1(\mathbf{k}), \mathbf{k}) \Big/ \left.\frac{\partial\varepsilon_1(\omega, \mathbf{k})}{\partial\omega}\right|_{\omega=\omega_1(\mathbf{k})}$$

for the group velocity, $\mathbf{u}_g$. The relation (3.37) is perhaps not quite as useful as one might think; it implicitly assumes that only *one* dispersion relation enters the problem. Unfortunately, this is not always so, in particular not in the kinetic theory for ion acoustic waves [39]. The problem can be outlined as follows: assume for the sake of argument that two relevant branches of a dispersion relation are found, one with a large group velocity u_g and a large value of the damping rate $\omega_2(\mathbf{k})$, and another branch with small u_g and small $\omega_2(\mathbf{k})$. For the *initial value problem* we should be concerned with the latter branch, and might tend to forget completely about the other one. However, if we should then calculate a *spatial* damping for a given fixed real frequency ω, the result would be in error because waves with the least *spatial* damping (small $\mathbf{k}_2$) might be associated with the other branch, all depending on the actual values of u_g and $\omega_2(\mathbf{k})$.

The results of this section are in principle applicable for any medium supporting electrostatic waves, but the physical mechanism giving the damping depends on the medium. For collisonless plasmas that have particular attention here, the mechanism is Landau damping [29]. For collisionless plasma (in the sense discussed before) the expression (3.12) can be used to give the approximate electron term in the limit relevant here

$$\varepsilon_2(\omega(k_1), k_1) = -\pi \frac{\omega_{pe}^2}{k_1^2} f_{0e}'\left(\frac{\omega_1}{k_1}\right), \quad (3.38)$$

to be used in (3.34) with the reference electron velocity distribution being $f_{0e}(u)$. The wave damping is found to be expressed by the slope of the electron velocity distribution function taken at the phase velocity ω_1/k_1 of the electron wave. In the limit of small damping for Langmuir plasma waves, the approximation $\omega_1 \approx \omega_{pe}$ might as well used. The physical mechanism giving rise to the linear Landau damping will not be addressed here. Accurate and careful discussions can be found in the literature [29, 30]. Unfortunately, also incorrect explanations can be found, invoking reflection of particles by a potential barrier. This is a nonlinear effect, and should not be mixed into a linear model used for describing incoherent scattering. These nonlinear effects have also been studied in great detail [54–56].

Expressions for ion sound waves can be found also. To obtain results for a small, finite ion temperature where $T_i \ll T_e$ we use as a first approximation $\varepsilon_1(k, \omega) \approx 1 + 1/(k\lambda_D)^2 - (\Omega_{pi}/\omega)^2$ and obtain

$$\omega_1(k) = k\sqrt{\frac{T_e/M}{1+(k\lambda_{De})^2}} \approx k\sqrt{\frac{T_e}{M}} \tag{3.39}$$

$$\omega_2(k) = \frac{\pi}{2}k\left(\frac{T_e/M}{1+(k\lambda_{De})^2}\right)^{3/2} f'_{0i}\left(\frac{\omega_1}{k}\right) \tag{3.40}$$

$$\approx \frac{\pi}{2}\left(\frac{T_e}{M}\right)^{3/2} k\, f'_{0i}\left(\sqrt{\frac{T_e}{M}}\right). \tag{3.41}$$

The limit of quasi-neutrality is obtained by letting $(k\lambda_{De})^2 \to 0$ with results given explicitly. In these latter limits, both $\omega_1(k)/k$ and $\omega_2(k)/k$ are found to be constants.

For Langmuir waves having phase velocities $\omega_1/k_1 \gg u_{the}$ it is safe to ignore the ion contribution to the wave damping as long as (3.38) can be used, since $f_{0i}(u = \omega_1/k_1)$ is negligible. For ion sound waves, a similar argument is no longer obvious. In the limit $k\lambda_{De} \ll 1$ for thermal plasmas the fluid expression $C_s = \sqrt{(T_e + \gamma T_i)/M}$ can be used as an estimate for the ion sound speed with γ being the ratio of specific heats. We have $f_{0e}(u = C_s)$ to be negligibly small compared to $f_{0i}(u = C_s)$. For thermal plasmas the ratio of the ion and electron contributions to the Landau damping of ion sound waves will depend on the temperature ratio T_e/T_i. Using a Hydrogen plasma as reference, it is found that for moderate temperature ratios the ion Landau damping will dominate, but in cases where $T_e/T_i \geq 25$ it will be the electrons contributing the most. When $T_i \approx T_e$, the ion sound waves are found to be heavily Landau damped [57] giving a large $\varepsilon_2(\omega, \mathbf{k})$. The analysis based on (3.33) will no longer apply for this case.

3.3.2 Mathematical Modeling of Collisions

The foregoing analysis assumed ideal collisionless plasmas. For physically realistic conditions with a finite plasma parameter, this assumption can not be exact, but it is often a sufficiently good approximation to allow the results to be applied. Unfortunately, there are relevant ionospheric regions where a collisionless model will fail, e.g., the ionospheric E-region [13, 49, 58]. The nature of the collisions will vary. In a fully ionized plasma the Coulomb collisions are the only ones relevant. As an estimate [30] for the collisional mean free path the expression $\ell_c \approx N_p \lambda_D$ can be used. For partially ionized plasma as found in the lower parts of the Earth's iono-sphere [59], the situation is much more complicated, in particular if collisions with molecules are important, distinguishing elastic and inelastic collisions. Understand-ing the physics of the collisional processes is a major challenge, but it is even more difficult to model their effects in the plasma dielectric functions. Several phenomeno-logical models have been proposed [30, 60–62] taking here as an example one of the simplest ones generalizing the Vlasov equation (3.7) to

$$\frac{\partial f}{\partial t} + \mathbf{u} \cdot \nabla f + \frac{\mathbf{K}}{m} \cdot \nabla_{\mathbf{u}} f = -\nu(f - n_0 f_0)$$

where the term on the right hand side contains a constant collision frequency ν giving a relaxation of the velocity distribution function to its average unperturbed value $n_0 f_0$, also here with a normalized $f_0(u)$. Linearizing this modified Vlasov equation it is found that in reality nothing is gained: with $f = n_0 f_0 + f_1$, the collision term becomes $-\nu f_1$. Introducing a modified velocity distribution function $f_1 \equiv f_* e^{-\nu t}$ we insert this into the linearized Vlasov equation and recover (3.7), but now for f_*, so the gain is only trivial. The problem with this oversimplified model is that the distribution relaxes to the average and not the local value. A slight improvement can be found [60] by a collision model $-\nu(f - f_0 n)$ which still relaxes the distribution function to f_0, but now to the local density n. Interestingly, this improved collision model allows some exact results to be found for the linearized problem [30, 63]. Choosing one of these collision models it is possible, at least in principle, to modify the dielectric function so that expressions like (3.6) can be applied. It has been demonstrated [62] that the collisions can have a significant effect on the shielding of charged particles.

All the standard phenomenological model equations accounting for collision have one flaw in common: the collision frequency enters as a deterministic constant. It is usually interpreted as a statistical average, but by this the random nature of the collisions is ignored. A solvable problem can be found [23], but only for a rather simple problem with non-interacting charges. The basic features of a model that allows for randomly occurring collisions is outlined in Appendix F.1.3.

Chapter 4
The Fluctuation-Dissipation Theorem

Abstract For fluctuations in thermal equilibrium we have the fluctuation-dissipation theorem as a fundamental and general result valid not only for plasma media. The analysis based on the dressed particle model should reproduce the results of this theorem when it is applied to plasmas with Maxwellian velocity distributions. With this test fulfilled, it is then justifiable to use the dressed particle model also for plasmas out of thermal equilibrium, although the plasma is required to be stable with respect to perturbations.

4.1 Relations to the Fluctuation-Dissipation Theorem

The dressed test particle model has a solid foundation in plasma physics but it might nonetheless be comforting to find an independent justification of the results. For plasmas in thermal equilibrium such a test can be found in the general fluctuation-dissipation theorem which is a cornerstone in classical statistical mechanics [22, 23, 64]. It has also been derived in quantum mechanics [65]. The theorem relates dissipation in a medium to the power spectrum of thermal fluctuations in equilibrium. Various aspects of the theorem have been extensively studied and some special cases are given individual names: Nyquist theorem, Johnson noise, Einstein relation, etc. Specifically, the fluctuation-dissipation theorem applies for small amplitudes. As stated explicitly, the model for the density fluctuations derived by the dressed particle approach relied on a linear model. Turbulent conditions are therefore not covered here since these are found for inherently nonlinear phenomena. To give a slightly more detailed discussion a reduced expression for the density power spectrum using (3.6) is applied, ignoring for simplicity the ion contribution. By $\mathbf{E} = -\widehat{\mathbf{k}} \, en/\varepsilon_0 k$ from Poisson's equation we derive the electric field power spectrum solely due to Langmuir oscillations and find the electric field energy density

© The Author(s), under exclusive license to Springer Nature Switzerland AG 2025
H. L. Pécseli, *Introduction to the Theory of Incoherent Scattering of Radar Waves from Plasmas*, SpringerBriefs in Physics, https://doi.org/10.1007/978-3-031-82652-8_4

$$\frac{1}{2}\varepsilon_0 \left\langle \mathbf{E}(\mathbf{k}, \omega) \cdot \mathbf{E}^*(\mathbf{k}, \omega) \right\rangle =$$

$$\frac{e^2 n_0}{2\varepsilon_0 k^2} \left| \frac{1}{\varepsilon_r^{(e)}(\mathbf{k}, \omega)} \right|^2 \iiint_{-\infty}^{\infty} \delta(\omega - \mathbf{k} \cdot \mathbf{u}) f_0^{(e)}(\mathbf{u}) d^3 u, \qquad (4.1)$$

where $\mathbf{E} = -i\mathbf{k}\phi$ was used with $\phi = -en/\varepsilon_0 k^2$ for electrostatic fluctuations by Poisson's equation with immobile ions. The product $-en$ is the charge density for the Langmuir waves.

Here it is an advantage to use a small "trick" [22] noting that the relation

$$\frac{1}{|a+ib|^2} = \frac{1}{a^2 - b^2}$$

can be written as

$$\frac{1}{|a+ib|^2} = \frac{1/(a+ib)}{a-ib} = \frac{Im\{1/(a+ib)\}}{Im\{a-ib\}} = -\frac{Im\{1/(a+ib)\}}{Im\{a+ib\}}$$

$$= -\frac{Im\{(a-ib)/(a^2-b^2)\}}{Im\{a+ib\}} = \frac{1}{a^2 - b^2},$$

noting the "minus"-sign, and $Im\{a - ib\} = -b$, with a and b both real. Using the form given before we can write

$$\frac{1}{\left| \varepsilon_r^{(e)}(\mathbf{k}, \omega) \right|^2} = -\frac{Im\left\{ 1/\varepsilon_r^{(e)}(\mathbf{k}, \omega) \right\}}{Im\left\{ \varepsilon_r^{(e)}(\mathbf{k}, \omega) \right\}}.$$

Use of the specific form for the plasma dielectric function (3.14) gives

$$Im\left\{ \varepsilon_r^{(e)}(\mathbf{k}, \omega) \right\} = \pi \frac{\omega_{pe}^2}{k^2} \iiint_{-\infty}^{\infty} \delta(\omega - \mathbf{k} \cdot \mathbf{u}) \mathbf{k} \cdot \nabla_{\mathbf{u}} f_0(\mathbf{u}) d^3 u,$$

and find

$$\frac{1}{2}\varepsilon_0 \left\langle \mathbf{E}(\mathbf{k}, \omega) \cdot \mathbf{E}^*(\mathbf{k}, \omega) \right\rangle =$$

$$-\frac{m}{2\pi} Im\left\{ \frac{1}{\varepsilon_r^{(e)}(\mathbf{k}, \omega)} \right\} \frac{\iiint_{-\infty}^{\infty} \delta(\omega - \mathbf{k} \cdot \mathbf{u}) f_0^{(e)}(\mathbf{u}) d^3 u}{\iiint_{-\infty}^{\infty} \delta(\omega - \mathbf{k} \cdot \mathbf{u}) \mathbf{k} \cdot \nabla_{\mathbf{u}} f_0^{(e)}(\mathbf{u}) d^3 u}. \qquad (4.2)$$

This form is general, but of limited use. However, for cases where $f_0(\mathbf{u}) = f_0(u)$ we have $\nabla_{\mathbf{u}} f_0(u) = \widehat{\mathbf{u}} f_0'(u)$ with $\widehat{\mathbf{u}}$ being a unit vector. The special case for a plasma in thermal equilibrium with a Maxwellian $f_0(u)$ has $\nabla_{\mathbf{u}} f_0(u) = -\mathbf{u}(m/T_e) f_0(u)$. This expression together with (4.2) and some properties of the δ-function from Appendix A gives

$$\frac{1}{2}\varepsilon_0 \langle \mathbf{E}(\mathbf{k}, \omega) \cdot \mathbf{E}^*(\mathbf{k}, \omega) \rangle = \frac{T_e}{2\pi\omega} Im \left\{ \frac{1}{\varepsilon_r^{(e)}(\mathbf{k}, \omega)} \right\}. \tag{4.3}$$

Integration of (4.3) with respect to ω gives

$$\frac{1}{2}\varepsilon_0 \langle \mathbf{E}(\mathbf{k}) \cdot \mathbf{E}^*(\mathbf{k}) \rangle = \frac{T_e}{2\pi} \int_{-\infty}^{\infty} \frac{1}{\omega} Im \left\{ \frac{1}{\varepsilon_r^{(e)}(\mathbf{k}, \omega)} \right\} d\omega$$

$$= \frac{T_e}{2} \frac{1}{1 + (k\lambda_{De})^2}, \tag{4.4}$$

where the ω-integral is found by contour integration. The electric field energy density is $\frac{1}{2}\varepsilon_0 \langle \mathbf{E}(\mathbf{k}) \cdot \mathbf{E}^*(\mathbf{k}) \rangle \approx T_e/2$ for small k, i.e., in equi-partition.[1]

It can be shown that the result (4.3) is identical to what would be found by the fluctuation-dissipation theorem when it is applied for plasma conditions in thermal equilibrium [22, 28]. For collision dominated plasma an assumption of local thermal equilibrium can usually be justified at all times to good accuracy, but it is possible to extend the analysis also to collisionless plasmas where the relaxation to thermal equilibrium is a slow process and the velocity distribution functions can deviate from Maxwellians. The dressed particle model can be applied when the velocity distribution functions of the plasma particles are known, or alternatively obtained by a fit to observations for stable systems out of thermal equilibrium. It is an interesting feature that classical thermodynamics assumes that thermal equilibrium is reached by irreversible dissipative processes. The Vlasov equation is a time-reversible equation, but the Landau damping it describes acts nonetheless as a dissipation mechanism consistent with the fluctuation-dissipation theorem.

---

[1] The form (4.3) with $\varepsilon_r^{(e)}(\mathbf{k}, \omega)$ given by (3.17), is valid for thermal equilibrium only (i.e., the electron temperature T_e appears explicitly), but it can be used as a guide for conditions close to equilibrium where T_e then represents an "effective" temperature. You might find the definition

$$\frac{T_{ef}}{M} \equiv -\frac{\iiint_{-\infty}^{\infty} \delta(\omega - \mathbf{k} \cdot \mathbf{u}) f_0(\mathbf{u}) d^3 u}{\iiint_{-\infty}^{\infty} \delta(\omega - \mathbf{k} \cdot \mathbf{u})(\mathbf{k}/\omega) \cdot \nabla_{\mathbf{u}} f_0(\mathbf{u}) d^3 u}$$

for an effective "fluctuation temperature" [22] for species with mass M and a velocity distribution function $f_0(\mathbf{u})$.

Chapter 5
Transition to Turbulence

Abstract The standard model for incoherent scattering assumes a linearized model to be applicable, i.e., it assumes small fluctuation levels. For plasmas close to linear instability, in particular plasma states that are the result of saturated linear instabilities, this linearized model no longer applies.

5.1 Transition to Turbulent Conditions

It is convenient to use the total electric field energy density as a reference. The integral of (4.4) with respect to wavenumbers, i.e., for all $\{k_x, k_y, k_z\}$, is diverging but it is hardly physically meaningful to include wavenumber regions where $|\mathbf{k}| > 1/\lambda_{De}$. With this restriction we find

$$\frac{1}{2}\varepsilon_0\left\langle|E|^2\right\rangle = \frac{T_e}{2}\iiint_{-1/\lambda_{De}}^{1/\lambda_{De}}\frac{1}{1+(k\lambda_{De})^2}d^3k \sim \frac{T_e}{2\lambda_{De}^3} = \frac{n_0 T_e}{2N_p}.$$

For hot dilute plasmas where the plasma parameter $N_p \gg 1$, the energy density in the fluctuating electric field is much smaller than the thermal energy density nT_e, and vanishes in the Vlasov limit, $N_p \to \infty$. It has been argued [27] that if any fluctuations are left in this limit, that would be what is called turbulence. It might not be easy to reach consensus on this point of view, but it can be seen as a minimum requirement.

It is evident that radar back-scatter can be observed irrespective of the plasma state: e.g., near thermal equilibrium or alternatively strongly turbulent. The question is how to describe that state of the plasma. A formal equation has been proposed [28] with a general structure showing how the energy in mode $\mathbf{k}$ evolves by nonlinear couplings

$$\frac{\partial}{\partial t}\left\langle E^2\right\rangle_\mathbf{k} + \left(\sum_{\mathbf{k}'} C(\mathbf{k},\mathbf{k}')\left\langle E^2\right\rangle_{\mathbf{k}'}\mathcal{T}_{C_{\mathbf{k},\mathbf{k}'}}\right)\left\langle E^2\right\rangle_\mathbf{k} + \gamma_{d\mathbf{k}}\left\langle E^2\right\rangle_\mathbf{k} =$$
$$\sum_{\substack{\mathbf{p},\mathbf{q}\\\mathbf{p}+\mathbf{q}=\mathbf{k}}} C(\mathbf{p},\mathbf{q})\left\langle E^2\right\rangle_\mathbf{p}\left\langle E^2\right\rangle_\mathbf{q}\mathcal{T}_{C_{\mathbf{p},\mathbf{q}}} + S_{D\mathbf{k}}\left\langle E^2\right\rangle_\mathbf{k}. \tag{5.1}$$

© The Author(s), under exclusive license to Springer Nature Switzerland AG 2025   43
H. L. Pécseli, *Introduction to the Theory of Incoherent Scattering of Radar Waves from Plasmas*, SpringerBriefs in Physics, https://doi.org/10.1007/978-3-031-82652-8_5

Here S_{Dk} is the discreteness source. For sufficiently large fluctuation levels the nonlinear terms on the right hand side will dominate the discreteness source for the fluctuations. This limit can be considered as a turbulent stage.

A strongly turbulent state is usually found for neutral fluids, i.e., one where many interacting degrees of freedom are excited at the same time. In plasmas on the other hand, turbulence is just as often often observed to be 'weak' [66]. For discussing this distinction, a nonlinear model wave equation [67–71] can be proposed in the form

$$\frac{\partial Q_\ell(\mathbf{k}, t)}{\partial t} + i\omega(\mathbf{k}) Q_\ell(\mathbf{k}, t) =$$
$$\sum_{m,n} \sum_{\mathbf{k}'} M_{\ell,m,n}(\mathbf{k}, \mathbf{k}') Q_m(\mathbf{k}', t) Q_n(\mathbf{k} - \mathbf{k}', t). \tag{5.2}$$

The indices $\{\ell, m, n\} \in \{x, y, z\}$ label components of the complex vector $\mathbf{Q}(\mathbf{k}, t)$. The nonlinear term in the right hand side contains the coupling coefficients $M_{\ell,m,n}$ between components for wavenumbers $\mathbf{k}$ and $\mathbf{k}'$. Any quadratically nonlinear partial differential equation with a first order time derivative can be brought in the form of Eq. (5.2) by a Fourier series representation of the spatial variables. One such example is the Navier-Stokes equation [72] where Q_ℓ represents the incompressible fluid velocity component u_ℓ, while $i\omega(\mathbf{k}) \to k^2 v$, where v is the fluid kinematic viscosity [73]. For plasma waves, on the other hand, Q_ℓ can represent one of the electric field components E_ℓ, while $\omega(\mathbf{k})$ is then a linear dispersion relation, which may be complex for some wave vector ranges to account for linearly unstable systems, or alternatively wave dampings in some other wave-vector ranges. Electromagnetic turbulent fields can also be relevant for plasmas, for Magneto Hydro Dynamic (MHD) turbulence [74], or for the more restrictive case of electron magneto hydrodynamics (EMHD) [75, 76].

Two different limits can be envisaged for equations like (5.2): one where the linear terms on the left side dominate the nonlinear term on the right hand side, and the case where it is the other way around. In the incompressible fluid case mentioned before it readily seen that for small k, the nonlinear term will dominate viscosity thus giving conditions for strong turbulence, entirely controlled by nonlinearity. For many (but not all) plasma conditions it is found that the linear term dominates nonlinearty and we find the limit of weak turbulence which has also bee extensively studied [34, 66, 77, 78]. The results have been tested by numerical simulations in specific limits, giving a convincing support [79].

Incompressible fluid turbulence is described by a quadratically nonlinear equation, the Navier-Stokes equation, which can form the basis for discussions also of plasma turbulence. For plasma media we can find also conditions that are described by cubically nonlinear models. A nonlinear Schrödinger equation is one example [34, 78, 80]. Turbulence modelling by these equations has also bee studied [81].

The source of free energy in plasma turbulence is usually a linearly unstable condition. This can, for instance, be due to an electron beam exciting high frequency electrostatic Langmuir waves which subsequently decay into low frequency wave-types [82]. For ionospheric conditions the free energy can be found in the interactions between the solar wind and the Earth's magnetosphere and also in tidal motions of the ionosphere [83].

Little can here be said in general concerning the saturated turbulent state in plasmas, in particular if magnetized conditions are considered. Collisionless plasmas can support nonlinear structures, BGK-modes [84, 85], and it is found that plasma instabilities can saturate into conditions where such modes are present.

Chapter 6
Conclusions

Abstract This section contains a short summary of the results from the analysis of incoherent scattering models using the standard results.

6.1 Conclusion

The dressed particle model is appealing by its simplicity and has the fortunate property that it can be tested by a different approach for systems in thermal equilibrium. In this case the fluctuation dissipation theorem [15, 22, 65] predicts the density fluctuations when the dissipative features of the plasma, i.e., its entire dielectric function is known. The advantage of the dressed particle model is that it can be applied also to (stable) plasmas out of thermal equilibrium. Magnetized plasmas have been covered only sparsely in the present notes. The analytical form based on expressions like (3.32) will be more complicated than the unmagnetized case, but the physical interpretations are similar, in particular what concerns the role of zeroes for the dielectric function.

The present notes consider only non-relativistic particles. This will suffice for most ionospheric applications. In fusion plasmas, in particular, we can have highly energetic particles, also very fast ions. The analysis has to be re-derived in order to account for these conditions [86].

The analysis explicitly assumed collisionless plasmas. Ion-electron collisions are of minor importance in the lower ionosphere, while collisions with neutrals can be important, in particular in the ionospheric E-region and below. Extension of the analysis that covers also this limit can be found in the literature [87, 88].

The present summary assumes incoherent scattering, i.e., from independent scatterers. They are here, as in other related works, modeled by randomly and independently moving dressed particles. It is evident that *any* assembly of electrons will scatter incoming electromagnetic waves. One might imagine a large scale electrostatic wave or wavepacket, being linear or nonlinear: this will also scatter radar waves but we might prefer to denote the process as *coherent scattering*. Most likely, such conditions have to be modeled independently, case by case.

H. L. Pécseli, *Introduction to the Theory of Incoherent Scattering of Radar Waves from Plasmas*, SpringerBriefs in Physics, https://doi.org/10.1007/978-3-031-82652-8_6

It might be imagined that conditions where the presence of charged dust particles is significant [89] can be included by trivial extensions of, e.g., (3.6). Indeed, if the dust particles can be seen as point particles with a fixed charge and at rest (or moving with a known velocity distribution) this generalization is straight forward, recalling that the electrons bound on dust must be compensated by a reduction in the density of free electrons. Such a model is, however, likely to be oversimplified. The charges on dust particles are fluctuating [90], and can depend on the particle velocity, in particular. When a dust particle is highly charged it might happen that one or more of the charging particles (i.e., electrons) may be "loosely bound" (having small binding energies) so they can oscillate with respect to some reference position without actually being removed from the dust. These properties of charged dust particles are not straight forward to model for incoherent scattering.

Negative ions can be present, relevant mostly in the D-region. Multiply charged aerosols of either sign of the charge can be important in the polar summer mesosphere [91]. In either of the cases mentioned here, two limits can be identified. We can have many particles in a Debye sphere and in this limit a continuum model will apply also for the dust. In the alternative case the aerosols or dust particles are dispersed in the plasma with their individual screening clouds. This limit requires a separate analysis [92].

A number of recent observations using incoherent radar scattering [93–95] have found data resembling those in Fig. 3.8, but under conditions incompatible with those used for deriving the results shown there. An overview of the data and their interpretation is given in the literature [96]. Eventually, a feasible explanation was found in nonlinear phenomena, decay of beam excited Langmuir waves decaying into another Langmuir wave and the observed ion acoustic wave [82] thereby giving an "overpopulation" of asymmetrically propagating ion sound waves different from what is found for a thermal spectrum. These naturally enhanced ion-acoustic lines, or "NEIALs", can not be accounted for by the simple dressed particle model that applies for linear wave phenomena only.

Appendix A
Dirac's δ-Function

A.1 Dirac's δ-Function, a Summary

Dirac's delta-function, $\delta(x)$, is used several times in these notes so an overview might be useful. Heuristically, the δ-function can be defined as one which is vanishing for all $|x| > 0$ and infinite for $x = 0$, in such a way that $\int_{-\infty}^{\infty} \delta(x)dx = 1$. It is surprising how far one can get with this simple definition, but it fails when it comes to discussing derivatives of the $\delta(x)$-function, for instance. The present appendix gives a summary of the properties of the δ-function in more detail, with finer mathematical aspects left to the literature [21, 23, 97–101]. From these introductory remarks it is evident that a δ-function only makes sense as an integrand, possibly multiplied with another function. A graph of $\delta(x)$ is, for instance, hardly feasible. The notion of a δ-function can be made meaningful in terms of *generalized functions*, "functions of functions" [100].

Introduce first the notion of an *acceptable function*, or "good function" in the sense of Lighthill [99]. Such an acceptable function is one which is everywhere differentiable any number of times in such a way that all its derivatives are $O(|x|^{-N})$ as $|x| \to \infty$ for all N.

- **Example:** For instance $\exp(-x^2)$ is an acceptable, or "good", function, while polynomials are not acceptable functions.

A sequence $f_n(x)$ of acceptable functions (in the present sense) is called regular if, for any acceptable function $F(x)$, we have that the limit

$$\lim_{n \to \infty} \int_{-\infty}^{\infty} f_n(x)F(x)dx$$

exists, implying in particular that the integral is finite for all n. Two such regular sequences are *equivalent* if, for all acceptable functions $F(x)$, this limit is the same for both sequences.

© The Editor(s) (if applicable) and The Author(s), under exclusive license
to Springer Nature Switzerland AG 2025
H. L. Pécseli, *Introduction to the Theory of Incoherent Scattering of Radar Waves from Plasmas*, SpringerBriefs in Physics, https://doi.org/10.1007/978-3-031-82652-8

- **Example:** Two sequences $\exp(-x^2/n^2)$ and $\exp(-x^4/n^4)$ are equivalent. For both of these two cases we have $\lim_{n\to\infty} \int_{-\infty}^{\infty} f_n(x)F(x)dx = \int_{-\infty}^{\infty} F(x)dx$.

A generalized function $f(x)$ is defined as a regular sequence of $f_n(x)$ of acceptable functions. Two generalized functions are said to be equal if the corresponding regular sequences are equivalent. Thus, each generalized function is a class of all regular sequences equivalent to a given regular sequence.

A generalized function $f(x)$ is said to be even or odd, respectively, if the integral $\int_{-\infty}^{\infty} f(x)F(x)dx = 0$ for all odd or even acceptable functions $F(x)$.

Using the notation introduced here, Dirac's $\delta(x)$-function can be defined by the sequence $f_n(x) = (n/\pi)^{1/2} \exp(-nx^2)$ and equivalent sequences which all have the property that $\lim_{n\to\infty} \int_{-\infty}^{\infty} f_n(x)F(x)dx = \int_{-\infty}^{\infty} \delta(x)F(x)dx = F(0)$. The δ-function is found to be be even in the sense discussed before.

Another representation [101] for a δ-function is

$$\delta(x) = \lim_{\gamma\to\infty} \frac{\sin(x\gamma)}{\pi x}, \tag{A.1}$$

see Fig. A.1. At $x \to 0$ the value of the expression (A.1) is γ/π. With increasing x it oscillates with a period $2\pi/\gamma$. The integral $\int_{-\infty}^{\infty} dx \sin(x\gamma)/\pi x = 1$, independent of γ. Therefore the limiting case of (A.1) as $\gamma \to \infty$ has all the required properties of a δ-function. The form (A.1) is special by taking both signs for any finite γ, but its limiting value is a positive function nontheless! Use of (A.1) proves an often used expression

$$\frac{1}{2\pi} \int_{-\infty}^{\infty} e^{ikx}dk = \delta(x), \tag{A.2}$$

since $\int_{-\infty}^{\infty} e^{ikx}dk$ can be considered as the limiting case $\lim_{L\to\infty} \int_{-L}^{L} e^{ikx}dk$. Then

$$\frac{1}{2\pi} \int_{-\infty}^{\infty} \cos(kx)dk = \delta(x) \qquad \text{while} \qquad \frac{1}{2\pi} \int_{-\infty}^{\infty} \sin(kx)dk = 0. \tag{A.3}$$

Admittedly, we are a bit "cavalier" here by taking the principal value of the integral without justification: depending on how the integration limits are prescribed, we can obtain anything in the range $\{-2; 2\}$ for $x \neq 0$. The divergence for $x \to 0$ remains.

As an alternative representation one might take

$$\delta(x) = \lim_{\alpha\to0} \frac{1}{\pi} \frac{\alpha}{\alpha^2 + x^2}, \tag{A.4}$$

which again fulfills the requirements posed by the basic definition of a δ-function.

Similarly we have

$$\delta(x) = \lim_{\alpha\to0} \frac{1}{\alpha\sqrt{\pi}} e^{-(x/\alpha)^2}, \tag{A.5}$$

Fig. A.1 Illustration of the sequence (A.1) as a function of x and γ

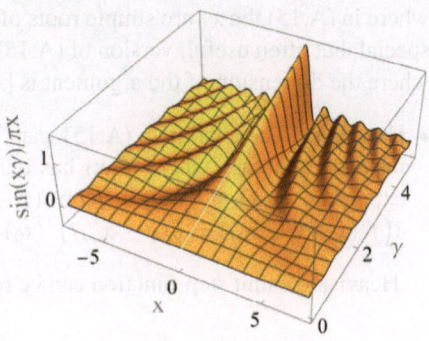

- **Example:** The Fourier transform of $\delta(x)$ is unity. The Fourier transform of $\delta(x - a)$ is its phase shifted value $\exp(ika)$.
- **Example:** The following relation is useful

$$\frac{1}{2\pi} \iint_{-\infty}^{\infty} \delta(x - ut)e^{-i(\omega t - kx)}dxdt = \delta(\omega - ku). \tag{A.6}$$

This is readily demonstrated.

The most important properties of one dimensional δ-functions can be summarized by the list

$$\int_{\alpha}^{\beta} F(x)\delta(x - a)dx = F(a) \quad \text{if} \quad \alpha < a < \beta, \tag{A.7}$$

$$\int_{\alpha}^{\beta} F(x)\delta(x - a)dx = 0 \quad \text{if} \quad a < \alpha < \beta \quad \text{or} \quad \alpha < \beta < a, \tag{A.8}$$

$$f(x)\delta(x - a) = f(a)\delta(x - a), \tag{A.9}$$

$$\delta(x) = \delta(-x), \tag{A.10}$$

$$x\delta(x) = 0, \tag{A.11}$$

$$\delta(ax) = \frac{1}{|a|}\delta(x), \tag{A.12}$$

$$\int_{\alpha}^{\beta} \delta(a - x)\delta(x - b)dx = \delta(a - b) \quad \text{if} \quad \alpha < a, \ b < \beta, \tag{A.13}$$

$$\delta(x^2 - a^2) = \frac{1}{2|a|}[\delta(x - a) + \delta(x + a)], \tag{A.14}$$

$$\delta(f(x)) = \sum_{j} \frac{\delta(x - x_j)}{\left|(df/dx)_{x=x_j}\right|}, \tag{A.15}$$

where in (A.15) the x_j are simple roots of $f(x) = 0$. The expression (A.14) is just a special, but often useful, version of (A.15). Note that the dimension of $\delta(x)$ is $[x]^{-1}$, where the dimension of the argument is $[x]$, this is easily realized.

- **Example:** The example (A.15) can be argued in detail by considering, for instance, the case where $f(x)$ has only one zero at x_0. Generally, by a Taylor expansion $f(x) \approx f(x_0) + f'(x_0)(x - x_0) = f'(x_0)(x - x_0)$. Then $\delta(f(x)) = \delta(f'(x_0)(x - x_0)) = \delta(x - x_0)/f'(x_0)$ by use of (A.12).

Heaviside's unit step function can be related to the δ-function by

$$\mathcal{H}(x) = \int_{-\infty}^{x} \delta(y)dy. \tag{A.16}$$

Dirac's δ-function can be directly related to the derivative of Heaviside's step function.

A.1.1 Derivatives of δ-Functions

Derivation of a δ-function can be defined from (A.1) as the limiting case of

$$\delta'(x) = \frac{1}{\pi} \lim_{\gamma \to \infty} \left[\frac{\gamma \cos(x\gamma)}{x} - \frac{\sin(x\gamma)}{x^2} \right]. \tag{A.17}$$

We find

$$\int_{\alpha}^{\beta} \delta'(x)F(x)dx = -F'(0),$$

with $\alpha < 0 < \beta$ and vanishing otherwise. A more general result for the n-th derivative $\delta^{(n)}(x)$ is $\int_{\alpha}^{\beta} \delta^{(n)}(x)F(x)dx = (-1)^n F^{(n)}(0)$. In particular, the first derivative of the δ-function satisfies the relation

$$x\delta'(x) = -\delta(x),$$

as easily demonstrated. Relations for $\delta''(x)$, etc., are also readily obtained, but rarely needed. The derivative of the δ-function is an odd function in the sense discussed before, $\delta'(-x) = -\delta'(x)$.

A.1.2 Multi Dimensional δ-Functions

The foregoing examples all assumed the δ-function to depend on one variable only. A "three dimensional" version can be defined as $\delta(\mathbf{r})$ which can be interpreted as

$\delta(\mathbf{r}) = (2\pi)^{-3} \int \exp(i\mathbf{k} \cdot \mathbf{r}) dk_x dk_y dk_z = \delta(x)\delta(y)\delta(z)$, having the basic property $\int \delta(\mathbf{r} - \mathbf{r}_0) F(\mathbf{r}) dx dy dz = F(\mathbf{r}_0)$. In particular, the dimension of $\delta(\mathbf{r})$ is $length^{-3}$.

Some useful three dimensional relations have no one-dimensional counterparts

$$\delta(\mathbf{r}) = \frac{1}{2\pi \, r^2} \delta(r),$$

and

$$\delta(\mathbf{r}' - \mathbf{r}) = \frac{2}{r^2} \delta(\widehat{\mathbf{q}}' - \widehat{\mathbf{q}}) \delta(r' - r),$$

where in the latter relation $\widehat{\mathbf{q}}'$ and $\widehat{\mathbf{q}}$ are unit vectors along $\mathbf{r}'$ and $\mathbf{r}$, respectively. An integration over r should here start at $r = 0$.

One important consequence of introducing a δ-function is that continuous and discrete charge distributions can be treated in formally the same way. In Poisson's equation, for instance, $\nabla \cdot \mathbf{D} = \rho(\mathbf{r}, t)$, a discrete charge distribution can be represented as $\sum_j q_j \delta(\mathbf{r} - \mathbf{R}_j(t))$, with $\mathbf{R}_j(t)$ being the positions, at time t, of charges q_j.

• **Example:** The equation

$$\nabla \cdot \mathbf{E} = \frac{a}{4\pi} \delta(\mathbf{r}),$$

has the solution $\mathbf{E} = a\widehat{\mathbf{r}}/r^2$ for all $|\mathbf{r}| > 0$, since for this case with spherical symmetry we have $\nabla \cdot (\widehat{\mathbf{r}}/r^2) = r^{-2}\partial(r^2\widehat{\mathbf{r}}/r^2)/\partial r = 0$ for all $|\mathbf{r}| > 0$. Using Gauss's divergence theorem $\int \nabla \cdot \mathbf{E} \, dV = \oint \mathbf{E} \cdot d\mathbf{s}$ on a spherical surface around the origin, we find $\int \nabla \cdot \widehat{\mathbf{r}}/r^2 dV = 4\pi$, completing the proof. Here, dV is the volume element in the integration.

Multi dimensional presentations [98] can be found for $\mathbf{r} = \{x_1, x_2, \ldots, x_n\}$ where

$$\delta(\mathbf{r}) = \delta(x_1)\delta(x_2)\ldots\delta(x_n).$$

with

$$\delta(\mathbf{r}) = \frac{1}{(2\pi)^n} \int_{-\infty}^{\infty} \ldots \int_{-\infty}^{\infty} e^{-i\mathbf{k}\cdot\mathbf{r}} d^n k.$$

A slightly more general form is

$$\delta(\mathbf{r}) = \frac{1}{(2\pi)^n} \lim_{\gamma \to 0} \int_{-\infty}^{\infty} \ldots \int_{-\infty}^{\infty} e^{-i\mathbf{k}\cdot\mathbf{r} - |\gamma| k^2} d^n k$$

$$= \lim_{\gamma \to 0} \left(\frac{1}{4\pi |\gamma|} \right)^{n/2} e^{-r^2/4|\gamma|},$$

where $r^2 = \sum_{j=1}^{n} x_j^2$. Another useful relation [98] is

$$\delta(\mathbf{r}) = \frac{1}{(2\pi)^{n/2}} \lim_{\alpha \to 0} \int_0^\infty e^{-\alpha k^2} \left(\frac{1}{kr}\right)^{(n-2)/2} J_{(n-2)/2}(rk) k^{n-1} dk$$

$$= \lim_{\alpha \to 0} \frac{e^{-r^2/4\alpha}}{(4\pi\alpha)^{n/2}},$$

where $\alpha > 0$, with J_n being a Bessel function. An alternative form can be found by

$$\delta(\mathbf{r}) = \frac{1}{(2\pi)^{n/2}} \lim_{\alpha \to 0} \int_0^\infty e^{-\alpha k} \left(\frac{1}{kr}\right)^{(n-2)/2} J_{(n-2)/2}(rk) k^{n-1} dk$$

$$= \lim_{\alpha \to 0} \Gamma\left(\frac{n+1}{2}\right) \frac{\alpha}{[\pi(r^2 + \alpha^2)]^{(n+1)/2}},$$

which can be seen as a generalization of (A.4).

Appendix B
Signal Modeling

B.1   Signal Modeling and Analysis

By a statistical average we usually mean ensemble averaging denoted by the brackets $\langle \rangle$. It is implicitly assumed that an ensemble of many (ideally infinitely many) realizations is available and the required average is performed over this. In many relevant cases a signal contains a deterministic element, a pulse-shape for instance, where some of the parameters entering happen to be statistically distributed.

In some cases it can be assumed that the system is ergodic, meaning that if a long (ideally infinitely long) realization (or several of these) is available, then the time averaging over one such realization will be equivalent to an ensemble averaging [102] no matter what quantity is averaged. This is a very strong statement!

It is sometimes (incorrectly) stated that time stationarity is sufficient for a system to be ergodic. An almost trivial counterexample [103] is an ensemble with each containing only a constant a_j that varies over the realizations of the ensemble as labeled by j. A time average (usually denoted by an "overline" or "overbar") clearly differs from the ensemble average, i.e., $\overline{a_j} \neq \langle a_j \rangle$. Time stationarity in the strict sense [23, 97, 103] assumes that all averages that can be imagined, such as $\langle f_j(t) f_j(t + \tau) \rangle$, etc. are independent of the absolute time t to depend only on time separations τ. Ergodicity implies for this example that $\langle f_j(t) f_j(t + \tau) \rangle = \overline{f_j(t) f_j(t + \tau)} \equiv \int_{-\infty}^{\infty} f_j(t) f_j(t + \tau) dt$ for any j. This point can be elaborated a little further: take a different representation where a Fourier transform is indicated by its variable as before

$$\langle f_j(t) f_j(t + \tau) \rangle = \left\langle \int_{-\infty}^{\infty} f_j(\omega) e^{i\omega t} d\omega \int_{-\infty}^{\infty} f_j(\gamma) e^{i\gamma(t+\tau)} d\gamma \right\rangle$$

$$= \int_{-\infty}^{\infty} \int_{-\infty}^{\infty} \langle f_j(\omega) f_j(\gamma) \rangle e^{i\omega t} e^{i\gamma(t+\tau)} d\gamma d\omega$$

© The Editor(s) (if applicable) and The Author(s), under exclusive license to Springer Nature Switzerland AG 2025
H. L. Pécseli, *Introduction to the Theory of Incoherent Scattering of Radar Waves from Plasmas*, SpringerBriefs in Physics, https://doi.org/10.1007/978-3-031-82652-8

where the rearrangement of terms is permissible since the integration paths are deterministic and unaffected by the ensemble averaging. To have this expression independent of t it is required that $e^{i\omega t}e^{i\gamma t}d\omega d\gamma \to \delta(\gamma + \omega)d\gamma d\omega$ giving

$$
\begin{aligned}
\langle f_j(t)f_j(t+\tau)\rangle &= \int_{-\infty}^{\infty}\int_{-\infty}^{\infty}\langle f_j(\omega)f_j(\gamma)\rangle e^{i\gamma\tau}\delta(\gamma+\omega)d\gamma d\omega \\
&= \int_{-\infty}^{\infty}\langle f_j(\omega)f_j(-\omega)\rangle e^{-i\omega\tau}d\omega \\
&= \int_{-\infty}^{\infty}\langle f_j(\omega)f_j^*(\omega)\rangle e^{-i\omega\tau}d\omega \equiv \int_{-\infty}^{\infty}\langle |f_j(\omega)|^2\rangle e^{-i\omega\tau}d\omega
\end{aligned}
$$

with $f_j(-\omega) = f_j^*(\omega)$. The term $\langle |f_j(\omega)|^2\rangle$ in the integrand is the power spectrum associated with the process $f(t)$. Since the ensemble and time averages are equivalent here, the spectrum is independent of the selected realization so that $\langle |f_j(\omega)|^2\rangle = \langle |f(\omega)|^2\rangle$.

It is generally believed that systems in thermal equilibrium are ergodic, but I know of no proof for this.

B.1.1 A Comment on Ergodicity

There is one final comment on ergodicity that need to be considered. As well known, any function $f(t)$ without singularities, and infinitely many times differentiable at all points can be Taylor expanded for all times t to have $f(t) = \sum_n^{\infty} f^{(n)}(t_0)(t-t_0)^n/n!$ for any choice of t_0, with $f^{(n)}(t_0) \equiv d^n f(t)/dt^n|_{t=t_0}$. For $n = 0$ the first term in the series is found to be $f(t_0)$, etc. The function $f(t)$ can contain parameters that vary over the ensemble but are constant in a given realization. These time series are deterministic in each of the realizations of the ensemble, and a time average in any one realization is thus not a *statistical* average! For a given time record to contain an element of randomness it is thus necessary that the Taylor expansion of $f(t)$ breaks down somehow, for instance by having singularities in one or more of the derivatives in one or more positions.

B.1.2 A Useful Model Problem: Synthetic Data

If the electrons are assumed to be randomly and independently distributed in an infinite space it might expected that the scattering from one electron will always be cancelled by scattering from some other one at another place. It is not so, and the simple model example analyzed in the present section illustrates this problem. Yes, sometimes an electron cancels the scattering from the reference electron, but sometimes another electron actually strengthens it. Constructive and destructive interference occurs randomly. With statistical problems you can not always trust your intuition!

It is a great advantage to have a simple model as a basis for discussions [104, 105]. Such a model should be based on some elementary statistical properties, and at least in principle allow relevant results to be obtained analytically [23, 97, 106, 107]. Consider first a simple example where a synthetic signal is generated by a random superposition of some basic structure $\psi(t)$. The model can for instance be used to illustrate the uncertainties associated with experimentally obtained averages when finite length records are used. With the present model a time record becomes

$$\Phi(t) = \sum_{k=1}^{N} a\psi(t - t_k). \tag{B.1}$$

The N positions labeled by $_k$ are randomly distributed in a large time interval $\mathcal{T}$. The pulses can overlap. In the following t_k is denoted "pulse arrival-time" because t is most often a time-like variable. This need not necessarily be the case, the model works for spatial variations as well. The distribution of t_k's is here taken to be uniform, meaning that the probability is $P(t_k) = 1/\mathcal{T}$ for $0 < t_k < \mathcal{T}$ and vanishing otherwise for all k. An amplitude a is included explicitly for later convenience. There are some constraints on permissible basic pulses $\psi(t)$ but these details need not concern us here.[1]

The time series (B.1) constitute a stationary random process and it will be demonstrated that for large $\mathcal{T}$ it is ergodic by construction. Examples from actual realizations are shown in Fig. B.1. The basic pulse was here taken to be $\psi(t) = \exp(-t^2)$ in all cases, with unit peak amplitude. In this case $\langle \Phi \rangle \neq 0$.

The model (B.1) have been analyzed in great detail [97, 106, 107]. It is general, and can refer to many physically relevant situations where some will be discussed in the following.

---

[1] Dealing with time-series like (B.1) there is a detail worth keeping in mind. If the basic pulse $\psi(t)$ is continuous and all its derivatives exist then a sum of such pulses will have the same properties, irrespective of the distribution of arrival times t_k. Then, by a Taylor expansion, the entire time series can, at least in principle, be predicted for all times and a statistical analysis of one selected realization can hardly be considered meaningful, see also Sect. B.1.1 The corresponding analysis of the entire ensemble of realizations of such time series with randomly distributed t_k will, on the other hand, remain meaningful. In order to bring an element of randomness into a selected realization of the ensemble it is necessary to have discontinuities either in the basic pulse or one or more of its derivatives. Examples can be $\psi(t) = ae^{-t}\mathcal{H}(t)$ or $\psi(t) = a\left(e^{-t}\mathcal{H}(t) + e^t\mathcal{H}(-t)\right)$, the first case being a discontinuous function, the second one continuous, but with discontinuous derivatives at $t = 0$. The analysis of one such case can be found in the literature [105, 108]. For the present summary, these details will have little consequence and are therefore ignored.

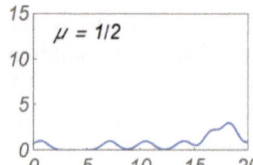 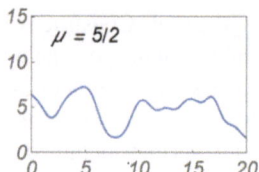 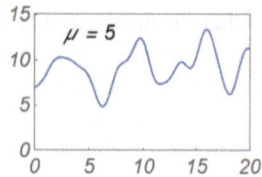

Fig. B.1 Examples of three realizations of the model (B.1) for densities $\mu = N/\mathcal{T} = 1/2$, $5/2$ and 5, respectively. Only a fraction of an extended data sequence is shown. Note the steady increase of the average level for increasing μ

B.1.3 Campbell's Theorem

Considering a time record as (B.1) a few basic results can be obtained. Campbell's theorem [23, 106, 109] thus states that the average of the signal is given by

$$\langle \Phi \rangle = \mu a \int_{-\infty}^{\infty} \psi(t)dt, \tag{B.2}$$

and the mean square value of the fluctuations about this average is

$$\langle (\Phi - \langle \Phi \rangle)^2 \rangle = \mu a^2 \int_{-\infty}^{\infty} \psi^2(t)dt, \tag{B.3}$$

where μ is the average number of basic structures $\psi(t)$ per time unit. The statement of the theorem is not precise until it is defined what is meant by "average". From the form of the equations one might be tempted to think of a time average; e.g. the value

$$\overline{\Phi(t)} \equiv \lim_{\mathcal{T} \to \infty} \frac{1}{\mathcal{T}} \int_0^{\mathcal{T}} \Phi(t)dt. \tag{B.4}$$

In the proof of the theorem, the average is generally taken over great many intervals of length $\mathcal{T}$ with t held constant. This is the ensemble averaging discussed before. To make the point more clear in the present context, the case of $\langle \Phi(t) \rangle$ can be taken for illustration. We observe $\Phi(t)$ for many, say M, intervals each of length $\mathcal{T}$, where $\mathcal{T}$ is large in comparison with the interval over which the effect $\psi(t)$ of a single basic structure, or pulse, is appreciable. Let ${}^n\Phi(t')$ be the value of $\Phi(t)$ at t' seconds after the beginning of the n-th interval indicated by the superscript ^{n}. Thus, t' is equal to t plus a constant depending upon the beginning time of the interval. A superscript is here placed in front because the usual place is reserved for another superscript later on. The value of $\langle \Phi(t') \rangle$ is then defined as

$$\langle \Phi(t') \rangle = \lim_{M \to \infty} \frac{1}{M} \left[{}^1\Phi(t') + {}^2\Phi(t') + \cdots + {}^M\Phi(t') \right], \tag{B.5}$$

assuming that this limit exists. The mean square value of the fluctuation of $\Phi(t')$ is defined in the very same way, and similarly for other related quantities.

As the expressions (B.2) and (B.3) of Campbell's theorem demonstrate, these averages as well as all the similar averages encountered later turn out to be independent of time. When this is true and when the M intervals in (B.5) are taken consecutively, the time average (B.4) and the average (B.5) become the same. To show that the time average equals the ensemble average, both sides of (B.5) are integrated from 0 to $\mathcal{T}$ and obtain

$$\langle \Phi(t') \rangle = \lim_{M \to \infty} \frac{1}{M\mathcal{T}} \sum_{m=1}^{M} \int_{0}^{\mathcal{T}} {}^{m}\Phi(t')dt'$$

$$= \lim_{M \to \infty} \frac{1}{M\mathcal{T}} \int_{0}^{M\mathcal{T}} \Phi(t)dt = \overline{\Phi(t)}. \tag{B.6}$$

This is the same as the time average (B.4), again assuming that the limit in (B.6) exists.

It is possible also here to choose the reference level so that $\langle \Phi(t) \rangle = 0$, but as the model construction clearly indicates, this will generally not be a particularly smart thing to do!

B.1.4   Proof of Campbell's Theorem

Consider first the case in which exactly K pulses are present in the record of duration $\mathcal{T}$. We think of these K pulses as determined to arrive in the interval $\{0, \mathcal{T}\}$ but any particular pulse is just as likely to arrive at one time as any other time, meaning that the pulse-arrival times are uniformly distributed in the interval $\{0, \mathcal{T}\}$. The pulses are numbered from one to K for purposes of identification, but the numbering need not have anything to do with the order of arrival. Thus, if t_k is the time of arrival of pulse number k, the probability that t_k lies in the interval $\{t, t + dt\}$ is $dt/\mathcal{T}$, irrespective of the arrival of any other pulse and independent of t. The number K will usually be different from one record to the next and is assumed to follow a Poisson distribution

$$P(K) = \frac{(\mu\mathcal{T})^K}{K!} e^{-\mu\mathcal{T}}. \tag{B.7}$$

The justifications and derivations for this distribution can be found in the literature [23, 106]. We have $\langle K \rangle = \sum_K K P(K) = \mu\mathcal{T}$ and $\langle K^2 \rangle - \langle K \rangle^2 = \langle K \rangle$ giving $\langle K^2 \rangle = (\mu\mathcal{T})^2 + \mu\mathcal{T}$ to be used later.

Take $\mathcal{T}$ to be very large compared with the range of values of t for which $\psi(t)$ is appreciably different from zero. In physical applications such a range usually exists and it is here denoted Δ even though it need not be very well defined. When exactly K pulses arrive in the interval $\{0, \mathcal{T}\}$ the effect is given by (B.1) where a subscript K

is included to emphasize that the number of pulses is fixed in the given interval $\mathcal{T}$. In other words; the intervals are *conditionally* selected with respect to K.

Suppose that a large number M of intervals of length $\mathcal{T}$ are examined. Out of these many intervals, the number having exactly K arrivals will to a first approximation be $M P(K)$ for the given K where $P(K)$ is the Poisson distribution (C.1). For a fixed value of t and for each interval having K arrivals, $\Phi_K(t)$ will have a definite value. As $M \to \infty$, the average value of the $\Phi_K(t)$'s, obtained by averaging over the intervals, is

$$\langle \Phi_K(t) \rangle = \int_0^{\mathcal{T}} \frac{dt_1}{\mathcal{T}} \cdots \int_0^{\mathcal{T}} \frac{dt_K}{\mathcal{T}} \sum_{k=1}^{K} a\psi(t - t_k)$$

$$= \sum_{k=1}^{K} \int_0^{\mathcal{T}} \frac{dt_k}{\mathcal{T}} a\psi(t - t_k). \tag{B.8}$$

If $\Delta < t < \mathcal{T} - \Delta$, we have effectively for large $\mathcal{T}$

$$\langle \Phi_K(t) \rangle = \frac{K}{\mathcal{T}} \int_{-\infty}^{\infty} a\psi(t)dt . \tag{B.9}$$

Averaging $\Phi(t)$ over all of the M intervals instead of only over those having K arrivals, we get by Bayes' rule, as $M \to \infty$,

$$\langle \Phi(t) \rangle = \sum_{K=0}^{\infty} \langle \Phi_K(t) \rangle P(K)$$

$$= \sum_{K=0}^{\infty} \frac{K}{\mathcal{T}} \frac{(\mu\mathcal{T})^K}{K!} e^{-\mu\mathcal{T}} \int_{-\infty}^{\infty} a\psi(t)dt$$

$$= \mu a \int_{-\infty}^{\infty} \psi(t)dt, \tag{B.10}$$

introducing the explicit form of the Poisson distribution $P(K)$ where also the density μ enters. Hereby the first part of Campbell's theorem is demonstrated. This rather elaborate proof on the relatively simple result (B.10) was used in order to illustrate a method which may be applied also for more complicated problems. Of course, (B.10) could be obtained by noting that the integral is the average value of the effect produced by one pulse arrival, the average being taken over one time unit, and that μ is the average number of arrivals per time unit. It is an almost trivial observation that even when $\langle \Phi(t) \rangle = 0$ an observer will usually find $\int_0^{\mathcal{T}} \frac{dt}{\mathcal{T}} \sum_{k=1}^{K} a\psi(t - t_k) \neq 0$ when considering one practically accessible realization of the ensemble with finite $\mathcal{T}$, even for large K.

In order to prove the second part, (B.3), of Campbell's theorem we calculate first $\langle \Phi^2(t) \rangle$. The definition of $\Phi_K(t)$ gives

$$\Phi_K^2(t) = \sum_{k=1}^{K}\sum_{m=1}^{K} a^2\psi(t-t_k)\psi(t-t_m).$$

Averaging this over all values of $t_1, t_2, \ldots, t_k$ with t held fixed as in (B.8) gives

$$\langle\Phi_K^2(t)\rangle = \sum_{k=1}^{K}\sum_{m=1}^{K} \int_0^{\mathcal{T}} \frac{dt_1}{\mathcal{T}} \cdots \int_0^{\mathcal{T}} \frac{dt_K}{\mathcal{T}} a^2\psi(t-t_k)\psi(t-t_m).$$

The multiple integral contains two principally different terms, those with $k = m$ and the others with $k \neq m$. If $k = m$ the integral value is

$$\int_0^{\mathcal{T}} a^2\psi^2(t-t_k)\frac{dt_k}{\mathcal{T}},$$

and if $k \neq m$ its value is

$$\int_0^{\mathcal{T}} a\psi(t-t_k)\frac{dt_k}{\mathcal{T}} \int_0^{\mathcal{T}} a\psi(t-t_m)\frac{dt_m}{\mathcal{T}}.$$

Counting the number of terms in the double sum it is observed that of the altogether K^2 terms there are K having the first value and $K^2 - K$ having the second. Hence, if again $\Delta < t < \mathcal{T} - \Delta$ we have

$$\langle\Phi_K^2(t)\rangle = \frac{K}{\mathcal{T}} \int_{-\infty}^{\infty} a^2\psi^2(t)dt + \frac{K(K-1)}{\mathcal{T}^2}\left(\int_{-\infty}^{\infty} a\psi(t)dt\right)^2.$$

Many relevant applications have $\langle\Phi(t)\rangle = 0$, thereby simplifying $\langle\Phi_K^2(t)\rangle$ as well as the following expressions where it might appear.

The foregoing analysis was conditional, i.e. it considered only realizations containing exactly K pulses. Averaging over all the realizations where K is statistically distributed we find, using that $\langle K(K-1)\rangle = \langle K\rangle^2$ with a Poisson distribution for K

$$\langle\Phi^2(t)\rangle = \sum_{K=0}^{\infty} P(K)\langle\Phi_K^2(t)\rangle$$

$$= \mu a^2 \int_{-\infty}^{\infty} \psi^2(t)dt + \langle\Phi(t)\rangle^2,$$

where the summation with respect to K is performed as in (B.10). After summation the value (B.10) for $\langle\Phi(t)\rangle$ is used. The coefficient μ in front of the integral is obtained even in case the distribution of K differs from a Poisson distribution.

The coefficient a can also be randomly distributed, so it can be different for different pulses. With this extension, the series become

$$\Phi(t) = \sum_{k=1}^{N} a_k \psi(t - t_k). \tag{B.11}$$

where the probability distribution for a_k is assumed known, with a_k being independent of the actual position of the pulse in the record. It is then easy to generalize the foregoing results to give, for instance

$$\langle \Phi^2(t) \rangle = \mu \langle a^2 \rangle \int_{-\infty}^{\infty} \psi^2(t) dt + \left(\mu \langle a \rangle \int_{-\infty}^{\infty} \psi(t) dt \right)^2. \tag{B.12}$$

Note that $\langle \Phi(t) \rangle$ scales linearly with μ. So does $\langle \Phi^2(t) \rangle$ when $\langle \Phi(t) \rangle = 0$. For the latter case this scaling does not seem obvious at all! A special case has $+a$ and $-a$ being equally probable. In that case $\langle \Phi(t) \rangle = 0$ while $\langle \Phi^2(t) \rangle > 0$.

A more general result containing also (B.12) is found by a derivation entirely similar to the one used before

$$R(\tau) \equiv \langle \Phi(t) \Phi(t + \tau) \rangle$$

$$= \mu \langle a^2 \rangle \int_{-\infty}^{\infty} \psi(t) \psi(t + \tau) dt + \left(\mu \langle a \rangle \int_{-\infty}^{\infty} \psi(t) dt \right)^2. \tag{B.13}$$

For time stationary random processes the *auto-correlation function* $R(\tau)$ is a function of time separations τ only. For non-stationary process more generally $R(t, \tau)$. By construction the time symmetry $R(-\tau) = R(\tau)$ is found for stationary processes. Evidently (B.13) contains (B.12) for the limit $\tau \to 0$. You might find [103] the definition *auto-covariance* for $C(\tau) \equiv R(\tau) - \langle \Phi(t) \rangle^2$.

The power-spectrum $G(\omega)$ for a random process is defined as the Fourier transform of the auto-correlation function, this is the Wiener-Khinchin theorem [103]. Often you find the normalization $C(\tau)/\langle \Phi^2(t) \rangle$ which, by construction, is unity for $\tau = 0$. The normalization implies that a time-scale constructed by integration

$$T_c \equiv \int_0^{\infty} \frac{C(\tau)}{\langle \Phi^2(t) \rangle} d\tau,$$

usually interpreted as a correlation time. If a time separation $\Delta t \gg T_c$ we can usually assume that the events at t and $t + \Delta t$ are uncorrelated (although you can easily find counterexamples, e.g. for oscillating correlation functions). Note that the lack of correlation is not the same as statistical independence, although it will be true for Gaussian random processes [103]. Nonetheless, you will find that uncorrelated signals will often be taken to be statistically independent without further ado.

The phase information in the basic pulse is lost when constructing the auto-correlation function. This is seen best by considering its Fourier transform, i.e., the power-spectrum. If we, for instance, change all phases by π, meaning that a change in the sign of the pulse, leaves the auto-correlation and the power-spectrum unchanged. Higher order correlations (triple correlations, etc.) and the corresponding spectra, the "bi-spectra", etc., retain this information, but these will not be need in the present summary.

The result (B.13) can be generalized [110] for two different but correlated signals giving the cross-correlation

$$R_{\Phi\Psi}(\tau) \equiv \langle \Phi(t)\Psi(t+\tau) \rangle. \tag{B.14}$$

For such cases two time series like (B.1) can be found where a structure or pulse $\phi(t - t_j)$ is placed in one record and a different structure $\psi(t - t_j)$ in the other one. The time symmetry of the auto-correlation function is lost in this case. The expression (B.13) is a special case of (B.14).

Appendix C
The Debye Length and Its Interpretations

C.1 Debye Shielding

The first ideas of incoherent radar scattering from the ionosphere assumed that this would be caused by a collection of independently moving electrons. This idea turned out to be at least partly incorrect: due to the long range Coulomb forces, an electron can interact with many others so that their motions become correlated. An important parameter for quantifying the importance of this collective interaction is the Debye length λ_{De} to be discussed in the following.

Physically, the Debye length is characterizing a shielding distance. A shielding can also be obtained when the ions are included in the analysis. The two cases, mobile and immobile ions, are often considered separately. It is important to emphasize that the following summary considers the re-adjustment of the plasma density when a point-like charge perturbation is introduced. We do *not* here consider losses of charged particles interacting with a solid surface: these effects are important when the body immersed in the plasma has a finite size. In two or three spatial dimensions there seems to be no problems with this: if the external charge is located at a point (in 3D) or a line (in 2D), the probability for plasma electrons and ions to reach physical contact with the reference charge can be ignored. We can, however, consider also a one-dimensional model, which physically corresponds to a charged plane. Admittedly, in this particular case it is a bit artificial to ignore physical contact between the external charge and the plasma particles. For such a model plasma particle losses are important. This and related cases should be be analyzed separately in relation to plasma probes [29, 111, 112].

C.1.1 Immobile Ions

Assume that a fixed point charge q is placed at the origin of the coordinate system and surrounded by a plasma with electron and ion densities n_e and n_i, respectively.

© The Editor(s) (if applicable) and The Author(s), under exclusive license
to Springer Nature Switzerland AG 2025
H. L. Pécseli, *Introduction to the Theory of Incoherent Scattering of Radar Waves from Plasmas*, SpringerBriefs in Physics, https://doi.org/10.1007/978-3-031-82652-8

Poisson's equation is written as

$$\nabla^2 \phi(\mathbf{r}) = \frac{e}{\varepsilon_0} \left(n_e(\mathbf{r}) - n_i(\mathbf{r}) - \frac{q}{e} \delta(\mathbf{r}) \right), \tag{C.1}$$

in terms of the electrostatic potential $\phi(\mathbf{r})$ with Dirac's $\delta(\mathbf{r})$-function introduced, see Appendix A. The electron charge is $-e$ with the present notation. Assume first that the ions with charge e are immobile and $n_i = n_0$. The problem here is time stationary and it is physically plausible to assume the electron distribution being of the form

$$f_e(\mathbf{u}, \mathbf{r}) = n_0 \left(\frac{m}{2\pi T_e} \right)^{3/2} \exp \left(- \frac{\frac{1}{2} m u^2 - e\phi(\mathbf{r})}{T_e} \right), \tag{C.2}$$

where T_e is a constant electron temperature. Since Boltzmann's constant always appears together with the temperature there is no need to keep it explicitly, so here and in the following it is included in T_e.

Integration of (C.2) with respect to velocity yields the isothermal Boltzmann distribution

$$\iiint_{-\infty}^{\infty} f_e(\mathbf{u}, \mathbf{r}) d\mathbf{u} \equiv n_e(\mathbf{r}) = n_0 \exp \left(\frac{e\phi(\mathbf{r})}{T_e} \right), \tag{C.3}$$

for the electrons in the potential $\phi(\mathbf{r})$.

C.1.2 Shielding in Three Spatial Dimensions with Spherical Symmetry

Unfortunately, it is not generally possible to solve (C.1) analytically for the electrostatic potential, $\phi(\mathbf{r})$. Expecting relevant magnitudes of the potential to be small, the approximation $e^x \approx 1 + x + \frac{1}{2} x^2 + \dots$ can used to linearize (C.3) giving

$$n_e(\mathbf{r}) = n_0 \left(1 + \frac{e\phi(\mathbf{r})}{T_e} \right), \tag{C.4}$$

and insert this approximation in (C.1). With immobile ions, $n_i = n_0$, the result is a simple closed equation for ϕ in the form

$$\nabla^2 \phi(\mathbf{r}) = \frac{e}{\varepsilon_0} \left(n_0 \frac{e\phi(\mathbf{r})}{T_e} - \frac{q}{e} \delta(\mathbf{r}) \right). \tag{C.5}$$

We can write Poisson's equation in spherical coordinates in terms of the corresponding ∇-operator. This will be appropriate when the disturbance of the plasma is caused by a point charge. With the given spatial symmetry the potential is varying with the radial variable only. The differential operator then becomes

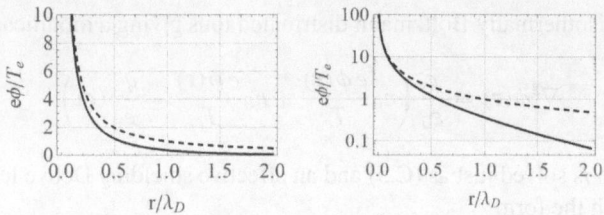

Fig. C.1 Illustration of the Debye shielding in 3 dimensions, assuming immobile ions. The dashed line is for the unshielded reference case

$$\nabla^2 \to \frac{1}{r^2}\frac{\partial}{\partial r}r^2\frac{\partial}{\partial r}.$$

Because of the given symmetry, n_e is also a function of the radial coordinate only and Poisson's equation is here readily solved to give

$$\phi(r) = \frac{q}{4\pi\varepsilon_0}\frac{\exp(-|r|/\lambda_{De})}{|r|}, \tag{C.6}$$

where the electron Debye length was introduced as

$$\lambda_{De} \equiv \sqrt{\frac{\varepsilon_0 T_e}{e^2 n_0}}. \tag{C.7}$$

See Fig. C.1 for an illustration in linear as well as in logarithmic presentation. It is readily seen that the Debye length is the distance an electron with mass m moving with a thermal velocity $\sim \sqrt{T_e/m}$ traverses in a time interval $1/\omega_{pe}$, with the electron plasma frequency being

$$\omega_{pe} \equiv \sqrt{\frac{e^2 n_0}{\varepsilon_0 m}}. \tag{C.8}$$

The reference potential was assumed to be $\phi(r \to \infty) = 0$. Close to the origin, i.e., close to the charge q, the solution has the form of the free-space solution $(q/4\pi\varepsilon_0)/|r|$, and a shielded potential for larger distances, with a shielding distance given by λ_{De}. For small distances the linearization used in (C.4) breaks down, but this is of little consequence since the plasma shielding is negligible there anyhow.

C.1.3 Mobile Ions

The model for Debye shielding assumed a fixed immobile test charge. Strictly speaking the assumption of immobile ions can then not be correct since they have sufficient time to adjust to the shielded electric fields. The model is readily extended by allowing for isothermally Boltzmann distributed ions giving a modification of (C.5) in the form

$$\nabla^2 \phi(\mathbf{r}) = \frac{e}{\varepsilon_0} \left(n_0 \frac{e\phi(\mathbf{r})}{T_e} + n_0 \frac{e\phi(\mathbf{r})}{T_i} - \frac{q}{e}\delta(\mathbf{r}) \right). \tag{C.9}$$

This equation is solved just as (C.5) and an effective shielding Debye length λ_{Def} is found through the form

$$\frac{1}{\lambda_{Def}^2} = \frac{1}{\lambda_{De}^2} + \frac{1}{\lambda_{Di}^2}, \tag{C.10}$$

with $\lambda_{Di} \equiv \sqrt{\varepsilon_0 T_i/e^2 n_0}$ being the ion Debye length. The expression (C.10) is readily extended to account for different ion species with different temperatures and possibly different charged states.

C.1.4   Moving Test Charges

The foregoing section considered shielding of stationary test charges. Intuitively, it might be expected that the results will be approximately the same if the particles are moving with a small velocity, well below the electron thermal velocity. A numerical study [113] demonstrated a modification of the shielding cloud already when the velocity became comparable to the ion sound speed. Analytical studies of the same problem are summarized in the literature [30]. For large particle velocities radiation of waves can be expected, reminiscent of the Cherenkov radiation. The problem will be analyzed in somewhat more detail in Appendix F.1. The fast particles will eventually decelerate due to the energy losses [22, 114], but this effect is not accounted for by the standard version of the dressed particle analysis.

C.1.5   The Plasma Parameter

From the Debye length and the plasma density a dimensionless number can be constructed as

$$N_p = n\lambda_{De}^3 \sim \frac{T_e^{3/2}}{\sqrt{n}}, \tag{C.11}$$

which apart from a factor of order unity is the number of particles in a sphere with radius λ_{De}. This number, called *the plasma parameter*, turns out to be important for classifying plasma conditions of interest. The definition of N_p is not unambiguous; sometimes the inverse of this number is called the plasma parameter, the proper choice will be evident from the context. For plasmas of interest we expect $N_p \gg 1$ with the definition used here. Note that N_p actually *decreases* for increasing plasma density with constant temperature, because the Debye length decreases as $\sim 1/\sqrt{n}$

for increasing density. Plasmas of interest (those with large N_p) will be *hot* and *dilute*, as for instance in the solar wind. As an exercise, estimate the plasma parameter of conduction electrons in copper at room temperature.

The plasma parameter can also be interpreted as a measure of the ratio of two length scales, the Debye length λ_{De} and an inter-particle separation $n^{-1/3}$. Large N_p implies that the average separation between particles, here ions and electrons, is much smaller than the Debye length. The so called "Vlasov limit" is reached when $N_p \to \infty$ is allowed. In this limit the particle discreteness of the plasma has disappeared and it forms an ideal continuum in phase space.

There are plasma conditions where the plasma parameter is not particularly large, but for ionospheric and magnetospheric applications we will almost always have $N_p \gg 1$. This means that many electrons and ions are interacting at the same time, giving rise to strong collective interactions. It tuns out that for these plasmas the collisional mean free path will be very large [29, 30], often we pretend it to be infinitely large and treat the plasma as a collisionless medium. Relaxation to thermal equilibrium will take a long time and the velocity distributions of electrons and ions can be very different from the Maxwellians characterizing thermal equilibrium. In particular, the temperatures of the two components (i.e., their mean kinetic energy densities) can be different, something often met in nature.

C.1.6 Details on the Number of Electrons in a Debye Sphere

The net amount of charge within a Debye sphere can be calculated. The density change induced by the charge q is first written as

$$n_e(r) - n_0 = \frac{q}{e} \frac{1}{4\pi \lambda_{De}^3} \frac{\exp(-r/\lambda_{De})}{r/\lambda_{De}}, \tag{C.12}$$

for the fully 3 dimensional problem considered here. Note that the dimensionality is important for the solution [30].

Integration of the charge density over all space gives

$$-e \int_0^{2\pi} \int_0^{2\pi} \int_0^\infty (n_e(r) - n_0) r^2 \sin\theta \, dr \, d\theta \, d\xi = -q \int_0^\infty \gamma \exp(-\gamma) d\gamma = -q.$$

The net induced charge in the Debye sphere is thus exactly what is needed to compensate the charge q which was introduced, a result which could have been guessed right away. Take the case where $q = e$. Then the net surplus of electrons in the Debye sphere is 1. For a sphere with radius λ_{De} placed at a randomly selected position in the plasma, the number N of electrons is a random variable and similarly for the ions. Let the electrons be placed independently of each other. This is not strictly correct [33], but the error in the assumption is small when the plasma parameter is large. With the given assumption, the number N will follow a Poisson

distribution and the fluctuations in the number of electrons in a given volume are
then $\sqrt{\langle (N - \langle N \rangle)^2 \rangle} = \sqrt{\langle N^2 \rangle - \langle N \rangle^2} = \sqrt{\langle N \rangle}$, where $\langle N \rangle$ is the average number
of electrons in that volume. Since the volume in question was taken to be the Debye
sphere, we have $\langle N \rangle = N_p \gg 1$. In terms of the plasma parameter $N_p = n\lambda_D^3$ the
fluctuations $\sqrt{N_p}$ in the number of electrons will generally be much larger than 1,
and the Debye sphere will have a somewhat hazy appearance [33]. If you could
see individual atoms you would hardly notice the electron cloud surrounding a test
charge, unless $q \gg e$. To observe the electron cloud you have to select an ion or
electron at rest and take an average over many realizations. To stay in the spirit of
Fig. 3.1 we show in Fig. C.2 a close-up of a wave source, serving as an illustration
of fluctuations in the near vicinity of a reference Debye-shielded particle.

C.1.7 Interaction Energy

To illustrate the physical meaning of the Debye length and the plasma parameter it
is worthwhile to consider some examples given in the following [30].

The induced potential, ϕ_{ind}, is defined as the difference between the actual potential in the plasma and eigen-potential, i.e., the $\sim 1/r$ potential variation associated
with the charge q in free space. The induced potential is in effect a measure of the
plasma's ability to shield the inserted charge q. Assuming for simplicity again the
ions to be immobile, the radial potential variation is found in the form

$$\phi_{ind}(r) = \frac{q}{4\pi\varepsilon_0} \frac{\exp(-r/\lambda_D) - 1}{r}. \tag{C.13}$$

The interaction energy, i.e., the potential energy of the charge q in the potential
induced in the plasma is then

$$q\,\phi_{ind}(r \to 0) = -\frac{q^2}{4\pi\varepsilon_0} \frac{1}{\lambda_D}, \tag{C.14}$$

since the charge will not have any potential energy in its eigen-potential. In obtaining
(C.14) the series expansion $e^{-x} = 1 - x + \frac{1}{2}x^2 \ldots$ was used.

Introducing the test charge q, the plasma density distribution is changed slightly, and thereby also the thermal energy distribution nT_e in the plasma. The perturbation in the thermal energy in a Debye sphere can be compared with the potential energy (C.14). Using (C.12) the change in thermal energy is found to be

$$\int_0^{2\pi} \int_0^{\pi} \int_0^{\infty} T_e(n_e(r) - n_0)r^2 \sin\theta dr d\theta d\xi = \frac{q}{e}T_e. \tag{C.15}$$

The ratio between the two energies (C.14) and (C.15) is

$$\left| \frac{q\,\phi_{ind}(r=0)}{T_e q/e} \right| = \frac{q^2 e}{4\pi\varepsilon_0 \lambda_{DQ} T_e}$$

$$= \left| \frac{q}{e} \right| \frac{1}{4\pi n \lambda_D^3} \equiv \left| \frac{q}{e} \right| \frac{1}{4\pi N_p}. \tag{C.16}$$

For plasmas of interest, where $N_p \gg 1$, the interaction potential energy is much less than the change in thermal energy in a Debye sphere, unless the charge q is very large. For instance, for charged dust particles very large values of q can be found, and in these cases the interaction potential energy can be important.

An important conclusion can be obtained from (C.16) by considering the interaction energy of a selected plasma particle in plasmas with $N_p \gg 1$. It can argueed that in order to have a large deflection of such a particle, its potential energy in the induced potential must be at least comparable to its kinetic energy. We learn from (C.16) that this is rarely so when $N_p \gg 1$, and expect that in this case collisions (meaning large deflections) are of minor importance. More detailed investigations [115] substantiate this argument.

C.1.8   Evacuation of a Debye Sphere

Assume that we want to evacuate all electrons from a small sphere with radius R. Considering again the ions as an immobile background with density n_0 it is argued that removing the first electron leaves one net positive charge behind, the next one two, etc. Each time an additional electron is removed we have to use an energy equivalent of what is needed to overcome the attraction by the surplus ions. To calculate the potential energy of the selected electron due to the ion charges we use Gauss' law with the radial component of the electric field $E_r = -d\phi/dr$

$$4\pi r^2 \frac{d\phi}{dr} = -\frac{4\pi}{3} r^3 \frac{e(n_0 - n_e)}{\varepsilon_0}, \tag{C.17}$$

for $r < R$, giving

$$\phi(r) = -\frac{e(n_0 - n_e(r))}{6\varepsilon_0} r^2 + \text{const.} \tag{C.18}$$

The energy needed to remove the last electron is calculated by taking the difference in potential energy from the center of the sphere at $r = 0$ to its edge at $r = R$ when $n_e(r \leq R) = 0$. Then

$$e\Delta\phi = -\frac{e^2 n_0}{6\varepsilon_0} R^2 .$$

Setting $e\Delta\phi$ equal to the thermal energy $\frac{3}{2}T_e$ of an electron, we can determine the radius R of that sphere which the electrons, at least in principle, can evacuate by their thermal energy. The result is $R = 3\lambda_{De}$. Thus, if for some reason all electrons in a localized region suddenly had their velocity vectors in the thermal motion point away from a fixed point, they would evacuate a small sphere with radius $\sim \lambda_{De}$. Of course without a Maxwell's demon this cannot happen in practice, but *energetically* it would be feasible. It seems safe to concluded that on scales comparable to or smaller than the Debye length, large fluctuations in electron density can be expected in thermal plasmas with a well defined electron temperature T_e.

The results of this subsection could also have been argued from (C.16), which simply states that the ratio of the interaction energy to the thermal energy for one electron is $1/(4\pi N_p)$. Since the Debye sphere contains approximately N_p electrons, the accumulated interaction energy of *all* the electrons is T_e, apart from a numerical constant not much different from unity.

The arguments presented before took the case with a sphere radius $R = \lambda_{De}$. Taking $R \gg \lambda_{De}$ instead it is, on the other hand, evident that by thermal motion the electrons will not be able to make any significant deviations from charge neutrality. On such *large* scales the plasma can be expected to be "quasi neutral," $n_e \approx n_i$, at least as long as we are dealing with phenomena in quasi-equilibrium with well defined plasma temperatures [30].

C.1.9   Quasi-Neutrality

This subsection quantifies the concept of quasi-neutrality, or "the plasma approximation" i.e., the approximation $n_e = n_i$. This cannot generally be true as an exact relation since it renders the right hand side of Poisson's equation identically zero. It was previously found that large deviations from charge neutrality *can* be found on scales comparable to or smaller than the Debye length. If, however, the analysis is restricted to scale lengths much larger than the Debye length and slow plasma variations, it is often safe to assume that no significant differences between n_e and n_i can occur; if by chance a large plasma volume lost even a small fraction of, say, the electrons, very large electric fields would be set up, which immediately would correct the imbalance. If the length scales are large, as assumed, this implies then that the gradients of the electric fields are small, even if the fields as such are nonzero. By the plasma approximation it can argued that Poisson's equation becomes redundant in the sense that, to the given approximation, it simply states $0 \approx 0$.

The arguments outlined here can be substantiated by considering Poisson's equation expressed in terms of ion and electron densities

$$\nabla \cdot \mathbf{E} = -\nabla^2 \phi = \frac{e}{\varepsilon_0}(n_i - n_e), \tag{C.19}$$

to be rewritten in normalized form. The normalized densities $\eta_e \equiv n_e/n_0$ and $\eta_i \equiv n_i/n_0$ are introduced, assuming a suitable normalizing reference density n_0, which is the same for electrons as well as ions. The electrostatic potential is normalized as $\psi \equiv e\phi/T_e$, whereby a local thermal equilibrium for the electrons is implicitly assumed. The arguments are therefore restricted to low frequency phenomena; only then it is possible to hope for the assumption of local electron thermal equilibrium to be accurate. Finally, it is assumed that the perturbations have a well defined scale length, $\mathcal{L}$. Writing Poisson's equation (C.19) in terms of new variables, with $\xi \equiv r/\mathcal{L}$, we readily find

$$\left(\frac{\lambda_{De}}{\mathcal{L}}\right)^2 \frac{\partial^2}{\partial \xi^2} \psi = \eta_e - \eta_i . \tag{C.20}$$

Assuming now that all the normalized variables ψ, η_i and η_e are of the same order of magnitude, while $\mathcal{L} \gg \lambda_{De}$, it can be argued that to lowest order accuracy we must have $\eta_i \approx \eta_e$ to have both sides of (C.20) equally small. (Note that by normalizing x by $\mathcal{L}$, it is ensured that $\partial^2 \psi/\partial \xi^2$ is also of order unity, when ψ is of the order unity.) This is the analytical basis for the quasi-neutral assumption. It allows us to consider the space-time varying bulk plasma density $n(\mathbf{r}, t)$, rather than the individual electron and ion densities. This can in some cases be a great simplification. The number of unknowns are reduced by one when setting $n_e \approx n_i \equiv n$ and the number of equations is also reduced by one when Poisson's equation is omitted.

In the quasi-neutral limit we have $\nabla^2_\xi \psi$ being equal to a product of a very small and a very large quantity, $\eta_i - \eta_e$ and $(\mathcal{L}/\lambda_{De})^2$, respectively. In that case we can let $\nabla^2_\xi \psi$ be unspecified by Poisson's equation, and to be determined by another equation. It is important to emphasize that the assumption of quasi neutrality does not impose any conditions on $\nabla^2 \phi$. The quasi-neutral assumption is not based on any direct assumption of linearization; it can be applicable also for nonlinear or large amplitude conditions.

Taking the two equations of continuity, one for electrons and one for ions, we assume quasi neutrality and find the density time derivatives to be $\partial n/\partial t$ in both equations. Taking the difference of these two equations gives $\nabla \cdot (n\mathbf{u}_e - n\mathbf{u}_i) = 0$. For singly charged ions this equation is multiplied by $-e$ to give divergence-free current densities, $\nabla \cdot \mathbf{j} = 0$ where $\mathbf{j} \equiv \mathbf{j}_e + \mathbf{j}_i$ is the sum of the current density contributions from the electron and ion motions. Ignoring Poisson's equation, it is no longer possible to obtain the full continuity equation for current from the continuity equations. For consistency with the quasi-neutral approximation, or the "plasma approximation," it is necessary to omit Maxwell's displacement current and use the original form of Ampere's law. Implicitly, this assumption implies again that $\omega E/c^2$ is negligible compared to $\mu_0 \mathbf{J}$. The assumption of quasi-neutrality has no association with the waves being electromagnetic or electrostatic! The assumption can be argued also without explicitly introducing the electrostatic potential as in (C.20).

The assumption of quasi neutrality has one more consequence which may not be evident from the derivation given here. When both Poisson's equation and Maxwell's displacement current no longer enter, there is no place in the basic equation where the permittivity of free space ε_0 enters. One consequence is that the electron as well as ion plasma frequencies have to be absent from the analysis. Quasi neutrality or the plasma approximation therefore applies only for very slow phenomena with frequencies $\omega \ll \Omega_{pi} \ll \omega_{pe}$. One way to see this is by noting that the quasi neutrality assumption does not involve the thermal velocities of neither ions nor electrons. From the basic definition of these quantities we have ions and electrons moving with their thermal velocities to traverse their respective Debye-lengths within their plasma periods, e.g., $\lambda_{Di}\Omega_{pi} = u_{thi}$ and similarly for electrons. Letting $\lambda_{Di} \to 0$ while keeping u_{thi} to remain constant, it is necessary to let $\Omega_{pi} \to \infty$ for consistency. Consequently all relevant frequencies must be low, $\omega \ll \Omega_{pi}$.

Appendix D
Elements of Electrostatics

D.1 Summary of Some Elements of Basic Electrostatics

In the following discussion of incoherent scattering the plasma waves entering the analysis will be electrostatic, i.e., the magnetic component will be negligible. Faraday's law states that $\nabla \times \mathbf{E} = -\partial \mathbf{B}/\partial t$, or for a plane wave $\mathbf{k} \times \mathbf{E} = \omega \mathbf{B}$. In order to have the magnetic field associated with the waves to be negligible it is require for consistency that $\mathbf{k} \times \mathbf{E} \approx 0$. To have a finite electric field it is then necessary to require $\mathbf{E} \parallel \mathbf{k}$. For electrostatic waves we have $\mathbf{E} = -\nabla \phi$ in terms of the electrostatic potential ϕ, consistent with the requirement $\mathbf{E} \parallel \mathbf{k}$.

D.1.1 Local Material Relations

The relation between electric fields and matter can be expressed in two different ways by two different versions of Poisson's equation. As an approximation take first the relative dielectric constant ε_r to be independent of position as well as frequency, making the results more general in a moment.

(1) the material relation relating electric fields to the dielectric displacement $\mathbf{D} = \varepsilon_0 \varepsilon_r \mathbf{E}$ where $\nabla \cdot \mathbf{D} = \zeta_{ext}$ in terms of the external charges $\zeta_{ext}(\mathbf{r}, t)$ controlled by the experimental conditions. All the internal material charges (i.e., all the electrons and ions constituting the material) are accounted for by the relative dielectric constant ε_r.

(2) As an alternative approach we have another version of the Poisson equation, $\nabla \cdot \varepsilon_0 \mathbf{E} = \zeta_{int} + \zeta_{ext}$, this time all charges indiscriminately added up on the right hand side. This version does not distinguish external and material charges and serves to determine the electrostatic electric field. Sometimes you might find the term "bound charges" for ζ_{int}, see Fig. D.1. For solid states this would be meaningful, but could be misleading for plasmas where, after all, the charges are

© The Editor(s) (if applicable) and The Author(s), under exclusive license to Springer Nature Switzerland AG 2025
H. L. Pécseli, *Introduction to the Theory of Incoherent Scattering of Radar Waves from Plasmas*, SpringerBriefs in Physics, https://doi.org/10.1007/978-3-031-82652-8

free to move with the sole constraint being that they participate in the collective plasma dynamics.

With $\mathbf{E} = -\nabla\phi$ we find from (1) that $\nabla^2\phi = -\zeta_{ext}/\varepsilon_0\varepsilon_r$. From (2) we have $\nabla^2\phi = -(\zeta_{ext} + \zeta_{int})/\varepsilon_0$. From this follows by eliminating ζ_{ext} that

$$\varepsilon_0\varepsilon_r = \varepsilon_0 + \frac{\zeta_{int}}{\nabla^2\phi}, \tag{D.1}$$

implying that ζ_{int} and $\nabla^2\phi$ are proportional for this simple case with $\varepsilon_r = \text{const.}$ An alternative form is

$$\zeta_{int} = \varepsilon_0(\varepsilon_r - 1)\nabla^2\phi.$$

More general material relations are needed to account for material dispersion, i.e., a frequency dependence of the response [23, 32, 53] as expressed by the Kramers-Kronig relations [32]. A constant dielectric response, independent of frequency, is after all found only in vacuum, although it is often used as an approximation in, for instance, a narrow frequency range.

D.1.2   Non-local Material Relations

More generally, ε_r can be taken to be a tensor in the case of anisotropic media. The electric displacement at a position $\mathbf{r}$ at a time t depends generally not only on the electric field at that position at that time, but also on $\mathbf{E}$ in neighboring positions with a spatially varying weight given through ε_r. In addition, the actual value of $\mathbf{D}$ depends in general also on the causal "history" of the electric field in the medium. These properties can be expressed through a Volterra type non-local relation, here given with a scalar ε_r rather than a tensorial form for simplicity

$$\mathbf{D}(\mathbf{r}, t) = \int_{-\infty}^{t}\iiint_{-\infty}^{\infty}\varepsilon_0\varepsilon_r(\mathbf{r} - \boldsymbol{\xi}, t - \tau)\mathbf{E}(\boldsymbol{\xi}, \tau)d^3\xi d\tau. \tag{D.2}$$

The time integration runs only up to the present time t. The expression (D.2) has the form of a convolution integral which simplifies a Fourier transformation [21]. The temporal variable requires particular attention here, eventually leading to the Kronig-Kramers relations [21, 32] between the real and imaginary parts of $\varepsilon_r(\mathbf{k}, \omega)$.

A Fourier transform of (D.2) with respect to the spatial and temporal variables gives $\mathbf{D}(\mathbf{k}, \omega) = \varepsilon_0\varepsilon_r(\mathbf{k}, \omega)\mathbf{E}(\mathbf{k}, \omega)$. With $\mathbf{E}(\mathbf{k}, \omega) = -i\mathbf{k}\phi(\mathbf{k}, \omega)$, Poisson's equation $\nabla \cdot \mathbf{D}(\mathbf{r}, t) = \zeta_{ext}(\mathbf{r}, t)$ is consequently expressed in the form

$$\varepsilon_0\varepsilon_r(\mathbf{k}, \omega)k^2\phi(\mathbf{k}, \omega) = \zeta_{ext}(\mathbf{k}, \omega),$$

together with

Fig. D.1 Illustration of polarization charges, without **a** and with **b** an externally imposed electric field

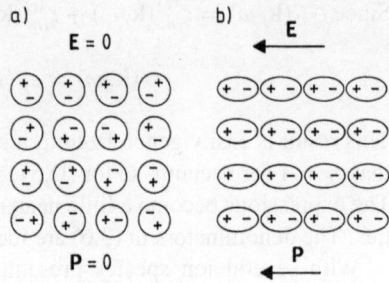

$$\varepsilon_0 \varepsilon_r(\mathbf{k}, \omega) = \varepsilon_0 - \frac{\zeta_{int}(\mathbf{k}, \omega)}{k^2 \phi(\mathbf{k}, \omega)}, \tag{D.3}$$

found by using $k^2 \phi(\mathbf{k}, \omega) = \left(\zeta_{ext}(\mathbf{k}, \omega) + \zeta_{int}(\mathbf{k}, \omega)\right)/\varepsilon_0$. An alternative form of the same result is

$$\zeta_{int}(\mathbf{k}, \omega) = \varepsilon_0 \left(1 - \varepsilon_r(\mathbf{k}, \omega)\right) k^2 \phi(\mathbf{k}, \omega). \tag{D.4}$$

A consistency check of (D.4) for vacuum, with $\varepsilon_r = 1$, gives $\zeta_{int}(\mathbf{k}, \omega) = 0$ as it should.

The dielectric function accounts for the medium properties and can be seen as a response function relating the electric field and the dielectric displacement [30, 116, 117]. To calculate an electrostatic electric field we take all charges, also those in the medium into account, while the dielectric displacement depends only on external (or "free") charges and the properties of the medium through $\nabla \cdot \mathbf{D}(\mathbf{r}, t) = \zeta_{ext}(\mathbf{r}, t)$. The external charges and associated currents are given entirely by $\zeta_{ext}(\mathbf{r}, t)$ and the continuity equation for charges. In order to have free electrostatic waves, i.e., some propagating in regions where local sources are absent, $\zeta_{ext}(\mathbf{r}, t) = 0$, it is necessary to have $\nabla \cdot \mathbf{D}(\mathbf{r}, t) = 0$ while $\mathbf{E}(\mathbf{r}, t) \neq 0$ there. This condition imposes constraints on ε_r in (D.2). This problem is analyzed best in a Fourier transformed presentation where the convolution integral in (D.2) has a simpler form.

D.1.3 Multi Species Plasmas

A medium, a plasma in particular, is often found to be composed of several species, here electrons and ions possibly several ion species. Dielectric responses for ions and electrons separately are introduced by

$$\varepsilon_r^{(e)}(\mathbf{k}, \omega) = 1 - \frac{\zeta_{int}^{(e)}(\mathbf{k}, \omega)}{\varepsilon_0 k^2 \phi(\mathbf{k}, \omega)} \quad \text{and} \quad \varepsilon_r^{(i)}(\mathbf{k}, \omega) = 1 - \frac{\zeta_{int}^{(i)}(\mathbf{k}, \omega)}{\varepsilon_0 k^2 \phi(\mathbf{k}, \omega)}$$

Since $\zeta_{int}(\mathbf{k}, \omega) = \zeta_{int}^{(e)}(\mathbf{k}, \omega) + \zeta_{int}^{(i)}(\mathbf{k}, \omega)$ we can use (D.3) to obtain a general form

$$\varepsilon_r(\mathbf{k}, \omega) = \varepsilon_r^{(e)}(\mathbf{k}, \omega) + \varepsilon_r^{(i)}(\mathbf{k}, \omega) - 1. \qquad (D.5)$$

This result is easily generalized to include multi-ion species, if needed [30]. Recall that $\varepsilon_r = 1$ for vacuum, so for (D.5) to include this case it is necessary to subtract 1. The expressions become a little neater if they are expressed in terms of susceptibilities. The denominators in (3.6) are identical to $\varepsilon_r(\mathbf{k}, \omega)$.

With several ion species present, having different masses, possibly different charges, etc., the expression (D.5) can be generalized to

$$\varepsilon_r(\mathbf{k}, \omega) = \varepsilon_r^{(e)}(\mathbf{k}, \omega) + \sum_s \left(\varepsilon_r^{(i,s)}(\mathbf{k}, \omega) - 1 \right). \qquad (D.6)$$

where the summation is over all ion species s. It is important to remember that each of the ion species are here assumed to be derived by a continuum model, the respective Vlasov equations. The individual plasma parameters have to be large for each the species [92]. The result (D.6) can also be expressed in terms of the susceptibility $\chi(\mathbf{k}, \omega) = \varepsilon_r(\mathbf{k}, \omega) - 1$.

Appendix E
The Complex Error Function

E.1 The Complex Error Function and Its Relation to the Plasma Dielectric Function

In deriving the dielectric function for a collisionless plasma we need a function, the so called Z-function

$$Z(z) \equiv \frac{1}{\sqrt{\pi}} \fint_{-\infty}^{\infty} \frac{e^{-x^2}}{x - z} \, dx, \tag{E.1}$$

that can be related to the complex error function which is known and tabulated [34, 36]. It can be nice to know how this relation can be obtained, but you can read the text without going into such details.

Using the integration contour from the center in Fig. 3.2, it is an advantage to introduce a new function that contains the Z-function as a special case

$$G(\alpha)e^{\alpha z^2} \equiv \fint_{-\infty}^{\infty} \frac{e^{-\alpha(x^2 - z^2)}}{x - z} \, dx, \tag{E.2}$$

and differentiate it with respect to α to find

$$\frac{d}{d\alpha} G(\alpha)e^{\alpha z^2} = - \fint_{-\infty}^{\infty} \frac{(x^2 - z^2)e^{-\alpha(x^2 - z^2)}}{x - z} \, dx$$

$$= -2ze^{\alpha z^2} \int_0^{\infty} e^{-\alpha x^2} dx = -ze^{\alpha z^2} \sqrt{\frac{\pi}{\alpha}}, \tag{E.3}$$

where the singularity in the denominator vanished. Integration of (E.3) gives

$$G(\alpha)e^{\alpha z^2} = -2z\sqrt{\pi} \int_0^{\alpha} e^{\alpha' z^2} \frac{d\alpha'}{\sqrt{\alpha'}} + i\pi$$

$$= -2z\sqrt{\pi} \int_0^{\sqrt{\alpha}} e^{(zt)^2} dt + i\pi$$

$$= -2\sqrt{\pi} \int_0^{z\sqrt{\alpha}} e^{t^2} dt + i\pi, \tag{E.4}$$

where we used $G(0) = i\pi$ by the given integration contour using the expression (E.2). All in all we have

$$G(\alpha) = -2e^{-\alpha z^2} \sqrt{\pi} \int_0^{z\sqrt{\alpha}} e^{t^2} dt + i\pi e^{-\alpha z^2}. \tag{E.5}$$

The last term in the integral $I(z)$ from (3.21) is now obtained by setting $\alpha = 1$ in (E.5), giving by (E.2) the equality

$$\fint_{-\infty}^{\infty} \frac{e^{-x^2}}{x - z} dx = -2e^{-z^2} \sqrt{\pi} \int_0^z e^{t^2} dt + i\pi e^{-z^2}.$$

The plasma dielectric function is then found in the form

$$\varepsilon(k, \omega) = 1 + \frac{1}{k^2 \lambda_{De}^2 \sqrt{\pi}} \left(1 - 2ze^{-z^2} \int_0^z e^{t^2} dt + iz\sqrt{\pi} e^{-z^2} \right).$$

with the previous definition of the variable $z = (\omega/k)\sqrt{m/2T_e}$.

We can introduce the plasma dispersion function [34, 36] for all complex z as

$$Z(z) = 2i \, e^{-z^2} \int_{-\infty}^{iz} e^{-t^2} dt$$

$$= 2i \, e^{-z^2} \left(\int_{-\infty}^{0} e^{-t^2} dt + \int_0^{iz} e^{-t^2} dt \right)$$

$$= i \, \sqrt{\pi} e^{-z^2} + 2i \, e^{-z^2} \int_0^{iz} e^{-t^2} dt$$

$$= -2e^{-z^2} \int_0^z e^{x^2} dx + i \, \sqrt{\pi} e^{-z^2} \tag{E.6}$$

using $\int_{-\infty}^{0} e^{-t^2} dt = \frac{1}{2}\sqrt{\pi}$, where also $t = i x$ was introduced one place. The relevance for $\varepsilon(k, \omega)$ becomes evident by comparing the two expressions. In (E.6) we recognize the error function with a complex argument.

Differentiation of $Z(z)$ gives the relation

$$Z'(z) = -2\left(1 + zZ(z) \right). \tag{E.7}$$

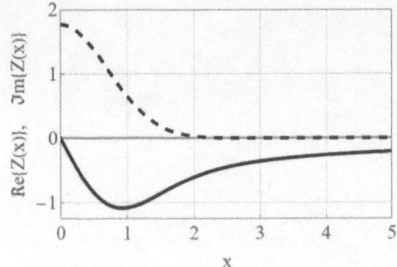

Fig. E.1 Real (full line) and imaginary part (dashed line) of the Z-function (E.1) for a real variable x. For $x = 0$ we have $Im\{Z\} = \sqrt{\pi} \approx 1.7725$ and $Re\{Z\} = 0$. The imaginary part $Im\{Z\}$ vanishes exponentially for $x \to \infty$, while the real part varies as $Re\{Z\} \approx -x^{-1}$

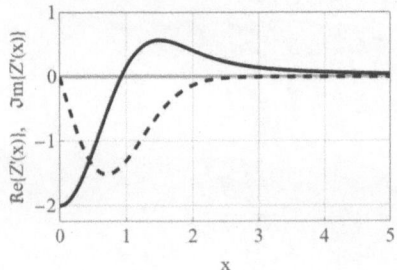

Fig. E.2 Real (full line) and imaginary part (dashed line) of the derivative Z' of the Z-function. For $x = 0$ we have $Im\{Z'\} = 0$ and $Re\{Z'\} = -2$. The imaginary part $Im\{Z'\}$ vanishes as the derivative of an exponential for $x \to \infty$, while the real part varies as $Re\{Z'\} \approx x^{-2}$

The expressions so far refer to the complex error function and its derivative as functions of a complex variable. When used in for instance (3.6) together with (3.14), we have the variables to be real (say x) and can write in terms of two real functions $Z(x) = Re\{Z(x)\} + i\ Im\{Z(x)\}$. These two functions are illustrated in Figs. E.1 and E.2. The expansion near the origin for real arguments, $x \ll 1$, are

$$Z(x) \approx -2x + 4x^3/3 + i\ \sqrt{\pi}(1 - x^2 + x^4/2)$$
$$Z'(x) \approx -2 + 4x^2 - 8x^4/3 - i\ 2\sqrt{\pi}(x - x^3), \tag{E.8}$$

while the asymptotic variations, $x \to \infty$, are

$$Z(x) \approx -x^{-1} - x^{-3}/2 - i\ 2\sqrt{\pi}\ \exp(-x^2)$$
$$Z'(x) \approx x^{-2} + 3x^{-4}/2 - i\ 4x\sqrt{\pi}\ \exp(-x^2). \tag{E.9}$$

The series expansions can be generalized to complex variables.

Appendix F
Illustrations of Dressed Particles

F.1 Charged Particle Dresses Described in Configuration Space

For reasons becoming evident in the following, it is not feasible to provide a complete or general kinetic model for the space-time variation of a charged particle dress in configuration space. Only two limiting cases will be discussed here, namely particle dress composed by electron waves in a fluid model and a corresponding simplified description of kinetic sound waves.

F.1.1 Fluid Model for Electron Plasma Waves

To illustrate the electron plasma waves (or Langmuir waves) trailing a fast moving charged particle first a relatively simple fluid model is presented here. In the limit of long wavelengths, and frequencies close to the electron plasma frequency, the deviations from a full kinetic model will not be significant. A general method for obtaining the wave radiation from a moving object can be found in the literature [30]. The analysis does not give an exact analytical result, but it is on the other hand sufficiently general to allow describing surface waves generated by a moving (small) boat, as well as from a fast charged particle moving in a plasma. Here we follow a more restrictive analysis, considering radiation of Langmuir waves only. These waves have a linear dispersion relation [29, 30, 118] in the form

$$\omega^2 = \omega_{pe}^2 + \frac{5}{3}u_{th}^2 k^2 = \omega_{pe}^2 \left(1 + \frac{5}{3}(k\lambda_{De})^2\right), \tag{F.1}$$

often called the Bohm–Gross dispersion relation [33, 52]. A dispersion relation like this is obtained by Fourier transforming a partial (or a set of partial) differential

© The Editor(s) (if applicable) and The Author(s), under exclusive license to Springer Nature Switzerland AG 2025
H. L. Pécseli, *Introduction to the Theory of Incoherent Scattering of Radar Waves from Plasmas*, SpringerBriefs in Physics, https://doi.org/10.1007/978-3-031-82652-8

equations, using $\partial/\partial t \leftrightarrow -i\omega$ and $\nabla \leftrightarrow i\mathbf{k}$. The transform is reversible, as stated by the symbol $\leftrightarrow$. A linear partial differential equation for the plasma dynamics can consequently be obtained from (F.1). It might be asked "dynamics of what", density, potential, velocity,...? It does not matter: the analysis is linear and any of the quantities can be derived from any of the others, e.g., the linear continuity equation gives $\partial\eta/\partial t = -\nabla \cdot \mathbf{u}$, etc. Making the reverse transformations mentioned before an equation for the normalized electron density η can be be obtained, for instance. The radiation of Langmuir waves by moving charges will be analyzed by the differential equation

$$\frac{\partial^2 \eta}{\partial t^2} + \omega_{pe}^2 \eta - \frac{5}{3} u_{th}^2 \nabla^2 \eta = 0. \tag{F.2}$$

In normalized units the equation can be written as $\partial^2\eta/\partial t^2 + \eta - \nabla^2\eta = 0$, where time is normalized by $1/\omega_{pe}$, velocities by u_{th} and lengths by u_{th}/ω_{pe}, being is the electron Debye length, apart from a numerical constant which is here assumed to be included in the definition of u_{th}. Changing the frame of reference to one moving with the particle, we find a stationary wave pattern. The replacement $\partial/\partial t \to -\mathbf{U} \cdot \nabla$, with $\mathbf{U} = U\hat{\mathbf{z}}$ being the test charge velocity gives

$$\left((U^2 - 1) \frac{\partial^2}{\partial z^2} + 1 - \frac{\partial^2}{\partial x^2} - \frac{\partial^2}{\partial y^2} \right) \eta = 0, \tag{F.3}$$

inviting a change in normalization of the z-coordinate to $z/\sqrt{U^2 - 1}$. By the construction it is seen that z becomes a "time-like" coordinate here.

This result is written for three spatial dimensions. For the case with rotational symmetry with respect to the z-axis we have $\nabla^2 \to r^{-1}\partial(r\partial/\partial r)/\partial r$ with $r = \sqrt{x^2 + y^2}$. For a two-dimensional case (corresponding to a moving thin charged wire), the term $\partial^2/\partial y^2$ is omitted in (F.3). In either case, the resulting equation has to be solved for $\eta = \delta(r)$, corresponding to a disturbance at the origin with analytical solutions available for some cases [118, 119]. Normalizing z by $\sqrt{U^2 - 1}$ the only requirement is found to be $U^2 > 1$, meaning that in order to give a radiation pattern, a charged particle has to move with a velocity exceeding the electron thermal velocity. On the other hand, a Debye shielding gives an adequate model for subthermal slow particles, $U^2 \ll u_{th}^2$.

The present results have shortcomings for $U^2 \sim u_{th}^2$ where electron Landau damping gives rise to significant modifications. The coefficient $5/3$ in (F.2) originates from a fluid analysis [30, 33, 52] using the ratio of the heat capacity at constant pressure to the heat capacity at constant volume, c_p/c_V. A full kinetic analysis gives a factor 3 instead. For $U^2 > 1$ the result will be representative since the numerical factor is contained in U anyhow.

The expression (F.3) is readily solved numerically, the result being very sensitive to the modeling of the source [30]. Taking $\partial^2\psi/\partial y^2 = 0$ for the 2 dimensional case mentioned before, the analytical solution [118] for a moving point source is

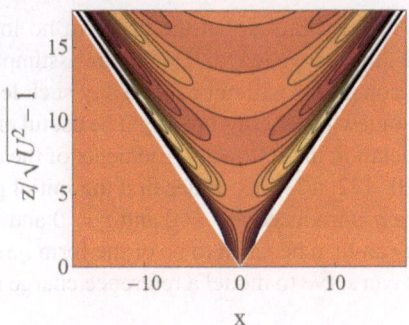

Fig. F.1 Example of a "dress" following a superthermal electron moving at normalized velocity U in a plasma with immobile ions. The illustration is based on a fluid model for high frequency electron (or Langmuir) waves [30] and does not account for Landau damping. By the scaling of the z-axis, the figure accounts for all velocities $|U| > 1$. The general form of the result is representative also for the spatially 3 dimensional case

$$\psi(x, z) = \frac{|x| \, J_1 \left(\sqrt{z^2 - x^2} \right)}{\sqrt{z^2 - x^2}} \quad \text{for} \quad z^2 > x^2, \tag{F.4}$$

for the two x-half-planes, in terms of the Bessel function J_1. The solution (F.4) is illustrated in Fig. F.1. The solution (F.4) is continuous and with continuous second derivatives for all $z^2 > x^2$.

F.1.2 Kinetic Model for Ion Acoustic Waves

A qualitative analysis for wave radiation by moving objects at arbitrary velocities [30] can be applied also for excitation of ion acoustic waves, but only by ignoring ion Landau damping. Assuming $T_e \gg T_i$, the ion acoustic dispersion relation [29, 120, 121] was used in the form

$$\omega^2 = \frac{C_s^2 k^2}{1 + (k\lambda_{De})^2}, \tag{F.5}$$

with $C_s^2 \equiv T_e/M$. For $k \to \infty$ there is a resonance at the ion plasma frequency $\Omega_{pi} = C_s/\lambda_{De}$. For $|k| \ll 1/\lambda_{De}$ we find the usual sound wave dispersion relation, $\omega^2 = C_s^2 k^2$. The previously mentioned method [30] gives good results only for high electron/ion temperature ratios [120]. These are unlikely to be found in the Earth's ionosphere, so results including the effects of ion Landau damping are needed here.

While the previously mentioned fluid analysis allows for deviations from quasi-neutrality in Poisson's equation, it turns out that a solvable kinetic model requires the assumption of quasi-neutrality (or the plasma approximation). One consequence of using a description in this limit is that the problem no longer has a built-in or natural length scale, i.e., the Debye length. Due to the assumption of point particles, the boundary or initial conditions will not contain any such length scale either.

The full analysis is somewhat lengthy, but might be useful also for other problems, so it is given in some detail in the following. Elements of the calculations are found also elsewhere [30, 120, 122, 123]. Consider first the initial perturbation to be that of a single point charge q introduced at $t = 0$ and $\mathbf{r} = 0$ and immediately removed again. The disturbance can then be taken to be of the form $q\delta(\mathbf{r})\delta(t)$. It will then be demonstrated how this can serve to model a reference charge moving with its dress.

Calculations in One Spatial Dimension

The relevant problem is here fully three dimensional and seems at first sight almost intractable. It turns out, however, that with some simplifying assumptions it is possible to deduce the requested result by starting from a one dimensional analysis. The argument is that a one dimensional slab geometry is after all a special case of a three dimensional analysis and it might be possible use this information to go from one result to the other [30]. Note that this transformation is possible only in selected problems and not universally. The analysis is here presented in detail, since the result is by no means obvious.

Consider first the one–dimensional version of the ion Vlasov equation. In its linearized form it becomes

$$\frac{\partial f}{\partial t} + u\frac{\partial f}{\partial x} - \frac{en_0}{M}\frac{\partial \phi}{\partial x} f_0'(u) = 0 \,. \tag{F.6}$$

The electrostatic field is introduced by $E = -\partial\phi/\partial x$ with $f(x, t, u)$ being the perturbation of the ion velocity distribution function, the unperturbed distribution being $n_0 f_0(u)$. The ions are here taken to be singly charged with $e > 0$. It is advantageous to normalize $\int_{-\infty}^{\infty} f_0(u)du = 1$, so the unperturbed reference density n_0 appears as a coefficient.

A second relation between f and ϕ is found by the Poisson's equation, containing also the externally imposed charge q which can have either sign

$$\nabla^2\phi = \frac{e}{\varepsilon_0}(n_e - n_i) - \frac{q}{\varepsilon_0}\delta(x)\delta(t)$$

$$= \frac{e}{\varepsilon_0}\left(n_0\frac{e\phi}{T_e} - \int_{-\infty}^{\infty} f(x, t, u)du\right) - \frac{q}{\varepsilon_0}\delta(x)\delta(t), \tag{F.7}$$

using the linearized expression for the relation between electron density perturbations and electrostatic potential imposed by the assumption of Boltzmann distributed electrons as in (C.4). The assumption of quasi-neutrality implies that the second derivative on the left side of (F.7) is negligible for scale-lengths of perturbations larger than the Debye length, given through the electron temperature T_e and n_0.

The Vlasov equation (F.6) is Fourier transformed in space and a Laplace transformed in time, using $-i\omega$ for the traditional Laplace variable s. This will simplify the notation. The relation (F.6) then becomes

$$-i\omega f(k, \omega, u) + iku f(k, \omega, u) - \frac{en_0}{M} ik\phi(k, \omega) f_0'(u) = 0, \qquad (\text{F.8})$$

using the initial condition $f(x, u, t = 0) = 0$. All perturbations are induced by the external charge q. With the assumption of Boltzmann distributed electrons and quasi-neutrality, the linearized Poisson's equation is replaced by

$$\int_{-\infty}^{\infty} f(k, \omega, u) du \equiv n(k, \omega) = n_0 \frac{e\phi(k, \omega)}{T_e} - \frac{q}{e}, \qquad (\text{F.9})$$

where $f(k, \omega, u)$ and $\phi(k, \omega)$ now denote the transformed functions. The Laplace and Fourier transforms of δ-functions are unity. Here, ω is a complex number with a positive imaginary part, while k is real. Eliminating $f(k, \omega, u)$ we find

$$n_0 \frac{e\phi(k, \omega)}{T_e} = \frac{q/e}{1 - \dfrac{T_e}{M} \displaystyle\int_{-\infty}^{\infty} \dfrac{f_0'(u)}{u - \omega/k} du}. \qquad (\text{F.10})$$

This equation can then be solved by performing the inverse Fourier and Laplace transforms. The inverse Fourier transform will be treated first. For simplicity the abbreviation $\psi \equiv n_0 e\phi/T_e$ is introduced giving

$$\psi(x, \omega) = \frac{1}{2\pi} \int_{-\infty}^{\infty} \psi(k, \omega) e^{ikx} dk. \qquad (\text{F.11})$$

In the following, we apply a method originally proposed by Mason [124], which has been used also elsewhere [30, 122, 125, 126]. In the integral in the denominator has the "standard" Landau singularity at $u = \omega/k$. When $k < 0$ the imaginary part of ω/k is negative and the integration path along the real u-axis is above the pole. Similarly, when $k > 0$ the integration path is below the pole. Therefore the integral (F.11) is split into two

$$\psi(x, \omega) = \frac{q/e}{2\pi} \left(\int_{-\infty}^{0} \psi_1(k, \omega) e^{ikx} dk + \int_{0}^{\infty} \psi_2(k, \omega) e^{ikx} dk \right) \qquad (\text{F.12})$$

where the two functions $\psi_{1,2}(k, \omega)$ are given by

$$\psi_{1,2}(k, \omega) = \frac{1}{1 - \dfrac{T_e}{M} \displaystyle\int_{1,2} \dfrac{f_0'(u)}{u - \omega/k} du}. \qquad (\text{F.13})$$

Fig. F.2 Integration
contours in the complex
k-plane. For the present
conditions the integrals
along the semi-circular parts
vanish as $R \to \infty$

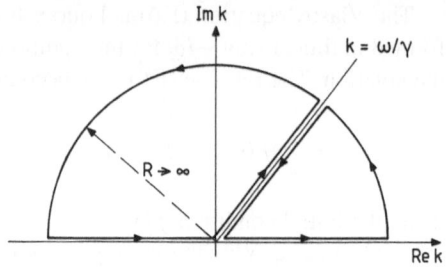

The two integration paths in (F.13) are different by running above and below the pole, respectively.

Given the line $\omega/k = \gamma$ in the complex k-plane as shown in Fig. F.2, it is found that in the region to the left of the line we have $\Im m\{\omega/k\} < 0$, i.e. the integration along the real u-axis runs above the pole and $\psi_1(k, \omega)$ is defined in this region. The notation $\Re e\{\}$ and $\Im m\{\}$ indicates real and imaginary parts. Similarly $\psi_2(k, \omega)$ is defined in the region to the right of the line in Fig. F.2. An integration can be performed around a closed contour in each region, where the results will depend only on the poles of the functions in (F.13). For stable ion conditions, these can only arise from the denominator becoming equal to zero. A Nyquist analysis [124] shows that for linearly stable plasmas the functions $\psi_{1,2}(k, \omega)$ are both analytical in the regions where they are defined. Thus there are no poles for the functions inside the integration contours and both closed contour integrals equal zero. For $R \to \infty$ in Fig. F.2, the integrals along the k-axis equal the integrals along the line $k = \omega/\gamma$ with the directions shown in Fig. F.2. The expression (F.12) then becomes

$$\psi(x, \omega) = \frac{q/e}{2\pi} \left(\int_\infty^0 \psi_1(\gamma) e^{i\omega x/\gamma} \left(\frac{-\omega}{\gamma^2} \right) d\gamma + \int_0^\infty \psi_2(\gamma) e^{i\omega x/\gamma} \left(\frac{-\omega}{\gamma^2} \right) d\gamma \right)$$

$$= \frac{q/e}{2\pi} \int_0^\infty \frac{\omega}{\gamma^2} \left(\psi_1(\gamma) - \psi_2(\gamma) \right) e^{i\omega x/\gamma} d\gamma. \tag{F.14}$$

The inverse Laplace transform is performed by noting that the inverse Laplace transform of $-i\omega e^{i\omega x/\gamma}$ is $-\delta'(x/\gamma - t)$ where $\delta'(x)$ is the derivative of Dirac's δ-function, see Appendix A. Then

$$\psi(x, t) = \frac{-iq/e}{2\pi} \int_0^\infty \frac{1}{\gamma^2} \left(\psi_1(\gamma) - \psi_2(\gamma) \right) \delta' \left(\frac{x}{\gamma} - t \right) d\gamma. \tag{F.15}$$

The substitution $z = x/\gamma$ gives

$$\psi(x, t) = \frac{iq/e}{2\pi x} \int_0^\infty \left(\psi_1\left(\frac{x}{z} \right) - \psi_2\left(\frac{x}{z} \right) \right) \delta'(z - t) dz.$$

Integration by parts leads to

$$\psi(x,t) = \frac{-iq/e}{2\pi} \int_0^\infty \frac{1}{z^2} \left(\psi_1'\left(\frac{x}{z}\right) - \psi_2'\left(\frac{x}{z}\right) \right) \delta(z-t)dz$$

$$= \frac{-iq/e}{2\pi} \left(\psi_1'\left(\frac{x}{t}\right) - \psi_2'\left(\frac{x}{t}\right) \right) \frac{1}{t^2}.$$

Using

$$\int_1 \frac{f_0'(\gamma)}{\gamma - x} d\gamma = \left(\int_2 \frac{f_0'(\gamma)}{\gamma - x} d\gamma \right)^*$$

for x real, with the asterisk denoting complex conjugate, we find by expressing the result in terms of ψ_2' with the integration paths given before

$$\psi(x,t) = \frac{q/e}{\pi t^2} h'\left(\frac{x}{t}\right) \tag{F.16}$$

where

$$h(\xi) = \Im m \left\{ \frac{1}{1 - \frac{T_e}{M} P \int_{-\infty}^\infty \frac{f_0'(u)}{u - \xi} du - i\pi \frac{T_e}{M} f_0'(\xi)} \right\}$$

$$= \frac{\pi \frac{T_e}{M} f_0'(\xi)}{\left(1 - \frac{T_e}{M} P \int_{-\infty}^\infty \frac{f_0'(u)}{u - \xi} du\right)^2 + \left(\pi \frac{T_e}{M} f_0'(\xi)\right)^2}.$$

We now have the plasma response to a perturbation of the form $q\delta(x)\delta(t)$. In the following this will be denoted $\psi_\delta^{(1)}(x,t)$. Obtaining illustrative results for arbitrary velocity distributions is not feasible. It will be assumed here that the unperturbed ion distribution $f_0(u)$ is well approximated by a Maxwellian, and the response can be expressed in terms of the plasma dispersion function. The integral in the denominator of $h(\xi)$ can then be evaluated explicitly. By making the substitutions $x = u\sqrt{M/2T_i} = u/u_{th}$ and $\gamma = \xi\sqrt{M/2T_i}$ with u_{th} being the ion thermal speed (here including a factor $\sqrt{2}$ for simplicity) we get

$$\int_{-\infty}^\infty \frac{f_0'(u)}{u - \xi} du = -\frac{M}{T_i}[1 + \gamma Z(\gamma)] = \frac{M}{2T_i} Z'(\gamma) \tag{F.17}$$

where $Z(\gamma)$ is the plasma dispersion function [36], see also Appendix E. Then

$$
h_M(\gamma) = -\Im m \left\{ \frac{1}{1 - \dfrac{T_e}{2T_i} Z'(\gamma)} \right\}
$$

$$
= \frac{\frac{1}{2} Q \, \Im m\{Z'(\gamma)\}}{\left(1 - \frac{1}{2}Q \, \Re e\{Z'(\gamma)\}\right)^2 + \left(\frac{1}{2}Q \, \Im m\{Z'(\gamma)\}\right)^2},
$$

where $Q \equiv T_e/T_i$. The subscript M is added on $h(\gamma)$ to emphasize the choice of a Maxwellian distribution. The present problem contains two velocity-scales, the ion sound speed $C_s = \sqrt{T_e/M}$ and the ion thermal velocity, their ratio being $\sqrt{Q}$. As mentioned already, there are no characteristic length or time scales here.

Calculations in Three Spatial Dimensions

The plasma response in three spatial dimensions can now be calculated by generalizing the one dimensional result. This is not always possible, but for the present case it can be done [30]. We will first postulate that the 3 dimensional response to a perturbation of form $q\delta(\mathbf{r})\delta(t)$ has the following form

$$
\psi_\delta^{(3)}(r, t) = \frac{q/e}{t^n} \mathcal{F}\left(\frac{r}{t}\right), \tag{F.18}
$$

where $\mathcal{F}$ is a function which is unspecified, for the time being.

The response in one dimension to a charged plane $\perp \hat{\mathbf{x}}$ can be calculated from the postulated expression (F.18) as

$$
\psi_\delta^{(1)}(x, t) = \frac{q/e}{t^n} \iiint_{-\infty}^{\infty} \delta(x') \mathcal{F}\left(\frac{\sqrt{(x-x')^2 + (y-y')^2 + (z-z')^2}}{t}\right) dx'\,dy'\,dz'
$$

$$
= \frac{q/e}{t^n} \iint_{-\infty}^{\infty} \mathcal{F}\left(\frac{\sqrt{x^2 + (y-y')^2 + (z-z')^2}}{t}\right) dy'\,dz'
$$

$$
= \frac{q/e}{t^n} \int_0^{\infty} \mathcal{F}\left(\frac{\sqrt{x^2 + \xi^2}}{t}\right) 2\pi\xi \, d\xi = \frac{\pi q/e}{t^n} \int_{x^2}^{\infty} \mathcal{F}\left(\frac{\sqrt{\gamma}}{t}\right) d\gamma,
$$

where first $\xi^2 = (y - y')^2 + (z - z')^2$ and later $\gamma^2 = \xi^2 + x^2$ was introduced. This is now the one–dimensional response to a δ-function in time and space which has to be equal to the result obtained in (F.16) giving

$$
\frac{\pi q/e}{t^n} \int_{x^2}^{\infty} \mathcal{F}\left(\frac{\sqrt{\gamma}}{t}\right) d\gamma = \frac{q/e}{\pi t^2} h'\left(\frac{x}{t}\right).
$$

Assuming dimensionless units, we get by differention with respect to x

$$\frac{1}{t^n}\mathcal{F}\left(\frac{x}{t}\right) = -\frac{1}{2\pi^2}\frac{1}{t^4}\left[\frac{t}{x}h''\left(\frac{x}{t}\right)\right]. \tag{F.19}$$

For consistency it must be so that $n = 4$ giving

$$\psi_\delta^{(3)}(r, t) = -\frac{q/e}{2\pi^2 t^4}\left(\frac{t}{r}\right)h''\left(\frac{r}{t}\right). \tag{F.20}$$

It is implicit in the arguments that the distribution function $f_0(u)$ is isotropic. Note the *self similarity* of this result; it depends on the ratio r/t rather than r and t separately, and is *scaled* by the factor t^{-4}, consistent with the initial assumption (F.18). On the other hand, it should also be noted that the self similarity of ψ in (F.16) is essential for obtaining a simple, closed form for the three dimensional response as derived from the one dimensional result (F.16). There are no general procedures to obtain a three dimensional space-time variation from a known one dimensional case.

The response to a charge moving along a path $z = U_0 t$ is now found by considering the moving charge as a continuous succession of δ-functions. Physically, this argument can be understood as being similar to generating the waveform in Fig. 3.1 by taking a small sack of pebbles and dropping them one-by-one in rapid succession along a straight line. When the time-interval between pebbles is small, the resulting waveform will approximate the one in Fig. 3.1. When superimposing many δ-function responses we might allow q to vary from one impulse to the next. The following model can be seen as a superposition of many charges inserted by infinitesimal displacements in space and time to represent one moving particle, taking q to be constant. The charge is here assumed move along the positive z-axis with speed U_0 giving

$$\psi(x, y, z, t) = -\frac{q/e}{2\pi^2}\int_0^t \frac{1}{(t - t')^4}\frac{t - t'}{\sqrt{x^2 + y^2 + (z - U_0 t')^2}}$$
$$\times h''\left(\frac{\sqrt{x^2 + y^2 + (z - U_0 t')^2}}{t - t'}\right)dt'. \tag{F.21}$$

This expression can be evaluated numerically when $h''_M(\gamma)$ is inserted. Since the analysis is linear, this result can be extended to a spatial distribution of many moving ions simply by superposition.

For a point charge, which has been moving with constant velocity for a long time, (F.21) can be simplified in the rest frame of the particle, taking the origin to be at $U_0 t$. The result is

$$\psi(x, y, z, t \to \infty) = -\frac{q/e}{2\pi^2} \int_0^\infty \frac{1}{\gamma^4} \frac{\gamma}{\sqrt{x^2 + y^2 + (z - U_0\gamma)^2}}$$
$$\times h''\left(\frac{\sqrt{x^2 + y^2 + (z - U_0\gamma)^2}}{\gamma}\right) d\gamma, \qquad (F.22)$$

using that $h''(\gamma)$ decays for large values of the argument. The density perturbation can be obtained from (F.22) to facilitate illustration in a plane which contains the straight-line trajectory of the particle. The disturbance induced by the moving charge propagates with velocity U_0, having a constant spatial shape. For thermal ions, the changes in radiation pattern caused by changes in electron-ion temperature ratios are illustrated in Fig. F.3. Changes in the temperature ratio imply changes in the ion sound speed and thereby also the Mach-angle. Related results for different test particle velocities are shown in Fig. F.4.

The foregoing summary gives an exact solution for the linearized ion Vlasov equation with the given assumptions. This is more general that anything obtained by the least Landau damped mode alone. It might be surprising to see a Cherenkov-like V-formation also for subsonic ions, but in a kinetic description of thermal ion dynamics the ion thermal velocity is also a characteristic velocity.

One aspect of the self-similarity can be emphasized: due to the assumption of quasi-neutrality, there is no natural length scale for the problem. Figures F.3 and F.4 use computational units for the axes. By reference to (F.22), we have that a change in units on the axes by a factor, say, β allows a change of integration variable to $\gamma' = \beta\gamma$, which ultimately results in a simple re-scaling of the response amplitude by a factor β^3. For a linear theory, as the present one, this amplitude scaling is immaterial, and only the spatial shape of the radiation pattern is relevant.

It is readily seen that for instance (F.22) diverges at $(x, y, z) \to 0$, by the divergence of $\int_0^\infty \gamma^{-4} d\gamma$, with $h''(U_0) \neq 0$. This is a consequence of the quasi-neutral assumption, which breaks down for separations comparable to or less than the Debye length as measured from the moving reference charge. For the same reason, the usual

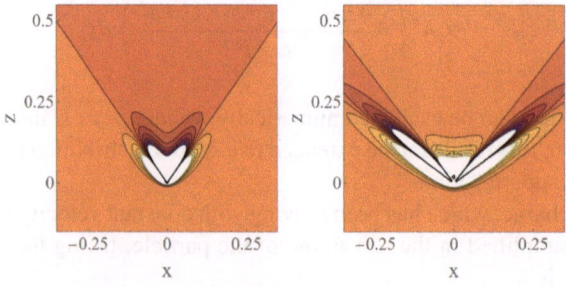

Fig. F.3 Radiation patterns for two temperature ratios, $T_e/T_i = 1$ and $T_e/T_i = 4$, both for test charge velocities $U_0 = 2.5\, u_{th}$. The ion Landau damping is strongest for the smallest temperature ratio. Since the ion sound speed increases with T_e/T_i and U_0 is the same for both figures, the first case is found to be near sonic, the second is subsonic

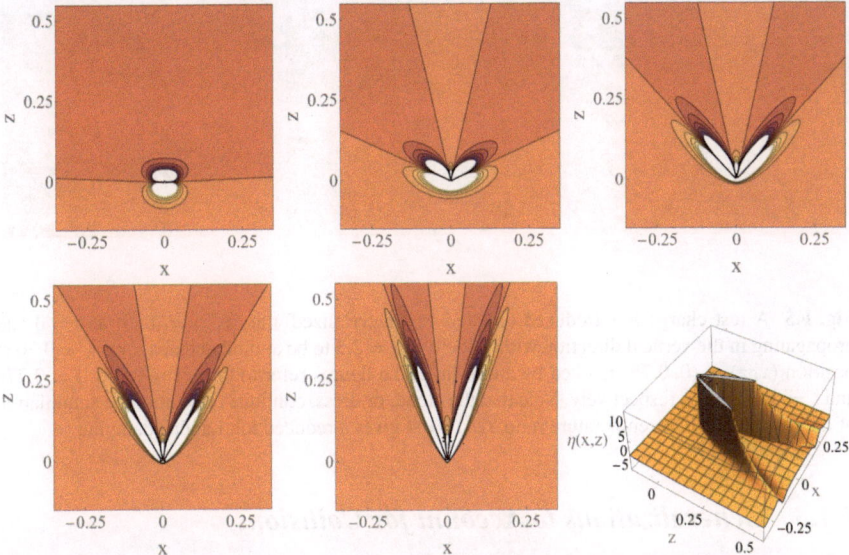

Fig. F.4 Plasma density response to a point charge moving with constant velocity U_0 in the plasma. The figure assuems $Q = 2$ and $U_0/u_{th} = 1, 1.75, 2.5, 3.25$ and 5. For $U_0/u_{th} = 5$ also a 3 dimensional surface plot is shown to help distinguish positive and negative parts of the density variation. There is a singularity at the origin $(x, z) = (0, 0)$ so the function has to be "cut off" at some level. All figures are shown in the moving frame of the test charge (placed in the origin), where the plasma is flowing in the z-direction. All these figures and those similar in the following are "cuts" in a a 3 dimensional variation. The full figure can here be obtained by a rotation with respect to the z-axis

"Airy-function" ripples [124] are not observed. These would be caused by acoustic wave dispersion, observed when the electron-ion temperature ratios are large. In a one dimensional treatment it is possible to avoid the quasi neutrality assumption and to retain the full Poisson equation, but the analytic result can not readily be used for obtaining the full three dimensional result. With the Debye lengths absent from the equations, there are no longer any characteristic quantities having the dimension "length" in the problem. With the assumption of a Maxwellian ion velocity distribution, the problem has the ion thermal velocity as a characteristic quantity, which was used for normalizing the self-similar variable x/t.

Figure F.4 shows examples for the ion Landau damped radiation pattern of ion acoustic waves induced by a moving test charge. The ion sound speed has no simple analytic expression for kinetic ion models. For a comparison we can use $C_s^2 \approx (T_e + 3T_i)/M$, which works best for large temperature ratios T_e/T_i. For moderate temperature ratios the approximation $C_s^2 \approx (T_e + T_i)/M$ is more appropriate. For neutral fluids, the transition from sub- to super-sonic conditions is abrupt and similarly for fluid models of ion acoustic waves [120]. For kinetic ion models the transition is more gradual. For $T_e = 2T_i$ it is approximately at $U_0 \approx 2.5u_{th}$.

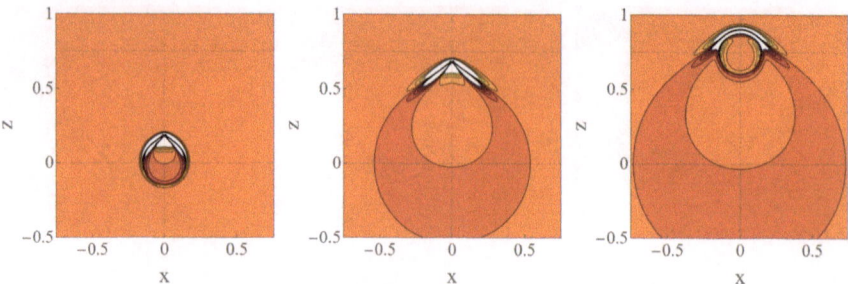

Fig. F.5 A test charge is introduced (created or "materialized") at $(x, z) = (0, 0)$ at $t = 0$ and propagating in the vertical direction with velocity $U_0 = 2.5$ to be annihilated at a time $t_a = 0.30$ at position $(x, z) = (0, 0.75)$ marked by a thin line. The figures refer to times $t = 0.075$, $t = 0.275$ and $t = 0.375 > t_a$, respectively. Note that the cloud, or dress, continues also after the annihilation of the test charge. The temperature ratio $T_e/T_i = 4$ gives a reduced ion Landau damping

F.1.3   Generalizations to Account for Collisions

Finally, we should like to point out that the results obtained here, notably (F.21), allows an illustration of a charge q manifested at a time t_1, which propagates along a straight line to be annihilated again at a time $t_2 > t_1$. First the screening cloud is being formed around the particle, then it propagates with an almost stationary cloud, which slowly dissolves by Landau damping when the charge q has disappeared. The larger the temperature ratio, the longer time will it take before the cloud disappears for $t > t_2$. This result can be used to construct the linearized model for two colliding charged particles. Before the collision, the two charges propagate along individual straight line orbits, and their responses are additive in the present linear model. The two (or more) trajectories can have any relative angle(s). At the collision time, the two (or more) charges are made to vanish, and at the same time new charges are introduced with their new respective velocity vectors so that charge neutrality as well as energy and momentum (for completely elastic collisions) are conserved at all instants. Momentum and energy conservation allows the new trajectories to be determined, and (F.21) can be used with $t = t_2$ referring to the collision time.

Figure F.5 illustrates an ion being created at $(x, z) = (0, 0)$ and propagating in the vertical direction, to be annihilated at a position $(x, z) = (0, 0.75)$ at a time $t = 0.3$ in dimensionless units. At $t \sim 0$ we find a "ring-like" emission of sound (a sphere in the fully three dimensional realization) while the moving ion is building up a wing-like dress, see also Fig. F.6. At the annihilation of the ion a ring-like sound emission is observed, while the dress generated by the ion prior to annihilation continues and slowly damps out. The intricate interaction between the decaying and newly formed particle dresses is illustrated in Fig. F.6. The collision model advocated before can be understood by taking the initial and final patterns in Fig. F.5 and superimposing them at the time and position of the selected collision. The reference particle changes its momentum in an infinitesimal time in the classical collision model where the duration

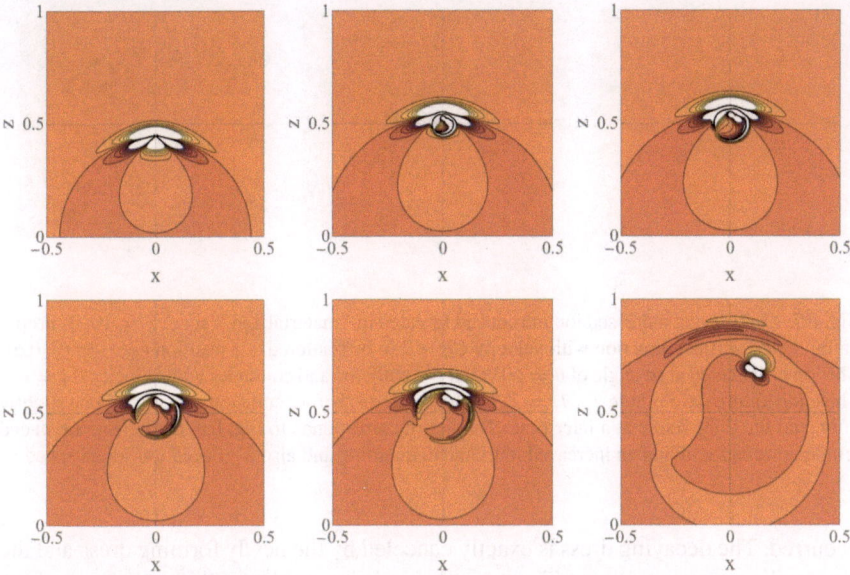

Fig. F.6 Time evolution of the cloud during collision of a dressed ion introduced (or "materialized") at $(x, z) = (0, 0)$ and propagating in the vertical direction with velocity $U_0 = 2$ to be scattered at a position $(x, z) = (0, 0.5)$. The ion is deflected at an angle $\theta = 60°$ at the collision to continue with velocity $U_0 = 1.2$. The temperature ratio is $T_e/T_i = 4$. With the time of the collision being at $t = 1$, we have the sequence of frames at times $t = 0.9, 1.1, 1.15, 1.2, 1.25$ and $1, 75$. These and similar figures still represent a cut in a fully 3 dimensional variation, but there is no longer any symmetry with respect to the z-axis, except for the first figure. The present figure can serve to illustrate a charge exchange collision where a neutral particle contributes to ensure energy and momentum conservation

of the collision is assumed to be negligible.[1] The cloud or dress will retain its memory and change slowly in comparison. To illustrate the dependence on the electron-ion temperature ratio results for $T_e/T_i = 2$ are shown in Fig. F.7. Here the ion Landau damping is enhanced as compared to the larger temperature ratio in Fig. F.6. The ion dress damps out in a short time when T_e/T_i is of the order of unity.

As a test, the analysis can be applied to a trivial case where the ion disappears at t_2 and the the new ion is at the same introduced at the same position with the same velocity and same direction of propagation. For this case, in effect, no collision has

---

[1] Except for hard sphere (or "billiard-ball") collisions, the duration of a collision can more generally be estimated as $\tau_c \approx b/u$, with u being the relative velocity between the two particles and b is the impact parameter [29]. Assuming an inverse power law dependence of the interaction force $F(r) \sim 1/r^s$ we find [22, 29] that the momentum change by a collision can be estimated as $\Delta(mu) \sim \tau_c F(b) \sim b/(ub^s)$. In order to have a significant momentum change by a collision it is assumed that $\Delta(mu) \sim mu$ and the corresponding impact parameter is then $b_c \sim u^{2/(1-s)}$. The corresponding collisional cross section becomes $\sigma \sim b_c^2 \sim u^{4/(1-s)}$. A finite duration of a collision necessitates velocity dependent collisional cross sections. In terms of a density of scatterers given as N, the collisonal mean free path is found to be $\ell_c \sim 1/(\sigma N) \sim u^{4/(s-1)}$.

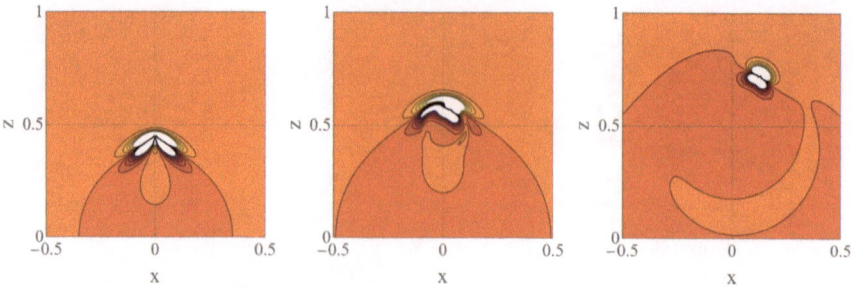

Fig. F.7 Collision of a dressed ion introduced (created or "materialized") at $(x, z) = (0, 0)$ propagating in the vertical direction with velocity $U_0 = 2$ to be scattered at a position $(x, z) = (0, 0.5)$. The ion is deflected at an angle of $\theta = 60°$ after the collision and continues with velocity $U_0 = 1.2$. The temperature ratio is here $T_e / T_i = 2$. The first frame is before collision, the second one slightly after, and the third frame at a later time. The figure corresponds to Fig. F.6, but here at a reduced temperature ratio, giving an increased ion Landau damping and also a reduced ion sound speed

occurred. The decaying dress is exactly canceled by the newly forming dress and the process has no consequence. The cancellation is incomplete if the parameters of the ion formed at t_2 differ from those of the one that decayed. In this case the interacting clouds, or dresses, will bear information of the collision.

Extending the model to, say, ion-ion collisions there would be 4 particles with their dresses entering: two incoming with their respective momenta, and 2 outgoing. Also such cases are readily modeled by superpositions like those illustrated in Figs. F.6 and F.7. The results are most easily comprehend when shown in a 2 dimensional or plane representation, and even then the results are not as transparent as those in the simpler collisions in Figs. F.6 and F.7. Figures for such more complicated collisions are not presented here.

The model has, so far, not provided details on the collision process. Collisional cross sections, their velocity dependence in particular, can be complicated for atoms and even more for molecular interactions. A vast amount of literature can be found, e.g., [127], covering mobilities of ions in a background of neutral gas. Collisions between an ions and a neutral particles where an electron is transferred (i.e., "charge exchange collisions" [58, 128]), change the ion to a neutral particle and the former neutral particle to an ion. Energetic ions will be cooled down when propagating through a cold neutral gas. These types collisions can be particularly important by having large cross sections and they have an important role also in the ionosphere. A brief summary of ionospheric collision frequencies can be found in the literature [59].

A few simple examples for the velocity dependence of collision cross sections are summarized here:

- Hard-sphere, or "billiard ball" collisions. Here the cross sections σ are constant giving the velocity dependence of the collision frequency as $\nu(u) \sim u$.

- Interactions by polarization forces. For so called "Maxwellian molecules" [129], we find for a wide velocity range $\sigma(u) \sim 1/u$ giving $v(u) \sim$ constant. This model is often used for illustrations of collisional processes.
- Coulomb interactions whave $\sigma(u) \sim 1/u^4$ giving $v(u) \sim 1/u^3$.

The velocity dependence of the collision frequency is found to be $v(u) \sim u/\ell_c \sim u^{(s-5)/(s-1)}$. It is readily demonstrated that the foregoing examples are special cases of this more general result [22].

Illustrations of a collision modeled by the foregoing procedure in configuration space are found in Figs. F.6 and F.7. The figures shown there can be interpreted as charge exchange collisions where a neutral particle looses an electron to a moving ion whereupon it continues as a charged ion while the previous ion continues as a neutral atom. Note the change in scales on the frames, compared to foregoing figures. Shortly after collision, the particle dress is strongly deformed: the new dress has to be formed, while the dress established prior to the collision has not been damped out yet. The main information of these results is found in noting that the collision is not affecting solely the colliding particle, but the entire cloud or dress moving with it.

A collision model can be proposed by the foregoing results. Assume that a probability density for a collision process is known. An often used model [23] for the probability of a collision in a small time interval dt being vdt with a constant v. In effect, it is assumed that the scattering cross section is inversely proportional to the relative velocity of the particles involved. The model is idealized, but not unphysical, applicable at least in some finite velocity interval, see e.g., the example of Maxwellian molecules mentioned before. The model gives the probability of no collision in a time interval $\{0 : t\}$ to be $P(0, t) = e^{-vt}$. More generally the Poisson distribution for K collisions in a time interval t is $P(K, t) = (vt)^K e^{-vt}/K!$. The average number of collisions in a time interval t is $\langle K \rangle = vt$. An illustrative collision model will consist of a random superposition of tracks like to one shown in Fig. F.7 with a length distribution determined by a collision model like the one outlined before. The velocity vector directions $\mathbf{U}_0$ are distributed over 4π in such a way that momentum is conserved. Each track in terminated by a collision, that forms the beginning of a new track, etc. More general numerical studies can be found in the literature [49].

The foregoing analysis was carried out for the electrostatic potential $\psi \equiv n_0 e\phi/T_e$. The assumed quasi neutrality in the present linearized model gives $\psi = n/n_0$, except near the position of the moving charge, so the results apply for the relative plasma density variations as well. The test charge q was treated as a "point particle" so that plasma particles are not absorbed upon direct collisions. This will happen for finite size charged objects, macroscopic dust particles, for instance.

As a summary a qualitative model can be suggested for the scattering in a weakly collisional plasma, i.e., one where the collision frequency is small compared to the ion plasma frequency. During its unperturbed orbit, a selected ion contributes to the scattering cross section as in a collisionless case. During a collision, a small localized burst of nearly spherically symmetric plasma waves are emitted as in Fig. F.6. These bursts contribute by a nearly isotropic radar scattering with a weak frequency dependence, almost like a δ-burst. It is found to be important that the presentation is carried out in configuration space rather than the Fourier domain used in e.g., (3.6). The latter case requires a calculation of a convolution of the free dressed particle and the Fourier transform of the window representing the time interval between two collisions. Formally the two presentations contain the same information, but it is comprehensible only in the space-time domain.

References

1. C. Fabry, Remarques sur la diffusion de la lumière et des ondes hertziennes par les electrons libres. C. r. hebd. Séance Acad. Sci. Paris, 187, 771–781 (1928)
2. W.E. Gordon, Incoherent scattering of radio waves by free electrons with applications to space exploration by radar. Proc. IRE 46, 1824–1829 (1958). https://doi.org/10.1109/JRPROC. 1958.286852
3. K.L. Bowles, Observation of vertical-incidence scatter from the ionosphere at 41 Mc/sec. Phys. Rev. Lett. 1, 454–455 (1958). https://doi.org/10.1103/PhysRevLett.1.454
4. E.M.A. Hussein, *Radiation Mechanics* (Elsevier Science, Principles and Practice, 2007). 978-0-08-045053-7
5. H. Salzmann, J. Bundgaard, A. Gadd, C. Gowers, K.B. Hansen, K. Hirsch, P. Nielsen, K. Reed, C. Schrödter, K. Weisberg, The LIDAR Thomson scattering diagnostic on JET (invited). Rev. Sci. Ins. 59, 1451–1456 (1988). https://doi.org/10.1063/1.1139686
6. R. Pasqualotto, P. Nielsen, C. Gowers, M. Beurskens, M. Kempenaars, T. Carlstrom, D. Johnson, JET-EFDA Contributors, High resolution Thomson scattering for Joint European Torus (JET). Rev. Sci. Ins. 75, 3891–3893 (2004). https://doi.org/10.1063/1.1787922
7. H. Bindslev, J.A. Hoekzema, J. Egedal, J.A. Fessey, T.P. Hughes, J.S. Machuzak, Fast-ion velocity distributions in JET measured by collective Thomson scattering. Phys. Rev. Lett. 83, 3206–3209 (1999). https://doi.org/10.1103/PhysRevLett.83.3206
8. J.V. Evans, Theory and practice of ionosphere study by Thomson scatter radar. Proceed. IEEE 57, 496–530 (1969). https://doi.org/10.1109/PROC.1969.7005
9. D.E. Evans, J. Katzenstein, Laser light scattering in laboratory plasmas. Rep. Progress in Phys. 32, 207 (1969). https://doi.org/10.1088/0034-4885/32/1/305
10. W.J.G. Beynon, P.J.S. Williams, Incoherent scatter of radio waves from the ionosphere. Rep. Progress in Phys. 41, 909 (1978). https://doi.org/10.1088/0034-4885/41/6/003
11. D. Alcaydé, editor. *Incoherent Scatter, Theory, Practice and Science*. Technical Report 97/53 - EISCAT Scientific Association, 1997. Collection of lectures given in Cargese, Corsica, 1995
12. D.H. Froula, S.H. Glenzer, Jr. N.C. Luhmann, J. Sheffield, *Plasma Scattering of Electromagnetic Radiation: Theory and Measurement Techniques* (Academic Press, USA, 2011)
13. E. Kudeki, M.A. Milla, Incoherent scatter spectral theories-part I: a general framework and results for small magnetic aspect angles. IEEE Trans. Geosci. Remote Sensing 49, 315–328 (2011). https://doi.org/10.1109/TGRS.2010.2057252
14. H. Akbari, A. Bhatt, C. La Hoz, J.L. Semeter, Incoherent scatter plasma lines: Observations and applications. Space Sci. Rev. 212, 249–294 (2017). https://doi.org/10.1007/s11214-017-0355-7
15. J.P. Dougherty, D.T. Farley, J.A. Ratcliffe, A theory of incoherent scattering of radio waves by a plasma. Proc. Royal Society of London. Series A. Math. Phys. Sci. 259(1296), 79–99 (1960)

© The Editor(s) (if applicable) and The Author(s), under exclusive license to Springer Nature Switzerland AG 2025
H. L. Pécseli, *Introduction to the Theory of Incoherent Scattering of Radar Waves from Plasmas*, SpringerBriefs in Physics, https://doi.org/10.1007/978-3-031-82652-8

16. J.A. Fejer, Scattering of radio waves by an ionized gas in thermal equilibrium. Canadian J. Phys. **38**, 1114–1133 (1960). https://doi.org/10.1139/p60-119

17. E.E. Salpeter, Electron density fluctuations in a plasma. Phys. Rev. **120**, 1528–1535 (1960). https://doi.org/10.1103/PhysRev.120.1528

18. M.N. Rosenbluth, N. Rostoker, Scattering of electromagnetic waves by a nonequilibrium plasma. Phys. Fluids **5**, 776–788 (1962). https://doi.org/10.1063/1.1724446

19. D.T. Farley, J.P. Dougherty, D.W. Barron. A theory of incoherent scattering of radio waves by a plasma II. Scattering in a magnetic field. Proc. Royal Soc. London. Series A. Math. Phys. Sci. 263, 238–258 (1961). https://doi.org/10.1098/rspa.1961.0158

20. T. Hagfors, Density fluctuations in a plasma in a magnetic field, with applications to the ionosphere. J. Geophys. Res. **1896–1977**(66), 1699–1712 (1961). https://doi.org/10.1029/JZ066i006p01699

21. D.C. Champeney, *Fourier Transforms and their Physical Applications* (Academic Press, London, 1973)

22. G. Bekefi, *Radiation Processes in Plasmas* (John Wiley and Sons, New York, 1966)

23. H.L. Pécseli, *Fluctuations in Physical Systems* (Cambridge University Press, Cambridge, UK, 2000)

24. W.B. Thompson, J. Hubbard, Long-range forces and the diffusion coefficients of a plasma. Rev. Mod. Phys. **32**, 714–718 (1960). https://doi.org/10.1103/RevModPhys.32.714

25. N. Rostoker, Superposition of dressed test particles. Phys. Fluids **7**, 479–490 (1964). https://doi.org/10.1063/1.1711227

26. R.L. Dewar, D. Leykam, Dressed test particles, oscillation centres and pseudo-orbits. Plasma Phys. Controlled Fusion **54**, 014002 (2012). https://doi.org/10.1088/0741-3335/54/1/014002

27. S. Ichimaru. *Basic Principles of Plasma Physics. A Statistical Approach.* Frontiers in Physics. Lecture Note Series. W. A. Benjamin, Inc., London, 1973

28. P.H. Diamond, S.-I. Itoh, K. Itoh, *Modern Plasma Physics*, vol. I (Physical Kinetics of Turbulent Plasmas. Cambridge University Press, Cambridge, UK, 2010)

29. F.F. Chen, *Introduction to Plasma Physics and Controlled Fusion*, 3rd edn. (Springer, Heidelberg, 2016)

30. H.L. Pécseli, *Waves and Oscillations in Plasmas*, 2nd edn. (Taylor & Francis, London, 2020), p.9781138591295

31. L. Chen, *Waves and Instabilities in Plasmas, World Scientific Lecure Notes in Physics*, vol. 12 (World Scientific, Singapore, 1987)

32. K.C. Yeh, C.H. Liu, *Theory of Ionospheric Waves*, volume 17 of *International Geophysics Series* (Academic Press, New York and London, 1972). ISBN: 0127704507

33. N.G. van Kampen, B.U. Felderhof, *Theoretical Methods in Plasma Physics* (North Holland Publishing Company, Amsterdam, 1967)

34. D.R. Nicholson, *Introduction to Plasma Theory* (John Wiley & Sons, New York, 1983)

35. J.K. Trulsen, N. Bjørnå, Influence of electrostatic electron waves on the incoherent scattering cross-section. Phys. Scripta **17**, 11–14 (1978). https://doi.org/10.1088/0031-8949/17/1/003

36. B.D. Fried, S.D. Conte, *The Plasma Dispersion Function* (Academic Press, New York, 1961)

37. E.C. Titchmarsh, *Introduction to the Theory of Fourier Integrals*, 2nd edn. (Clarendon Press, Oxford, 1948)

38. T.H. Stix, *Waves in Plasmas* (AIP, American Institute of Physics, New York, 1992)

39. R.W. Gould, Excitation of ion-acoustic waves. Phys. Rev. **136**, A991–A997 (1964). https://doi.org/10.1103/PhysRev.136.A991

40. J. Vierinen, B. Gustavsson, D.L. Hysell, M.P. Sulzer, P. Perillat, E. Kudeki, Radar observations of thermal plasma oscillations in the ionosphere. Geophys. Res. Lett. **44**, 5301–5307 (2017). https://doi.org/10.1002/2017GL073141

41. E. Marsch, S. Livi, Coulomb collision rates for self-similar and kappa distributions. Phys. Fluids **28**, 1379–1386 (1985). https://doi.org/10.1063/1.864971

42. W.-Z. Fu, L.-N. Hau, Vlasov-Maxwell equilibrium solutions for Harris sheet magnetic field with Kappa velocity distribution. Phys. Plasmas **12**, 070701 (2005). https://doi.org/10.1063/1.1941047

43. V. Pierrard, M. Lazar, Kappa distributions: theory and applications in space plasmas. Solar Phys. **267**, 153–174 (2010). https://doi.org/10.1007/s11207-010-9640-2

44. S. Ali, B. Eliasson, Slow test charge response in a dusty plasma with Kappa distributed electrons and ions. Phys. Scripta **92**, 084003 (2017). https://doi.org/10.1088/1402-4896/aa7c09

45. G. Livadiotis (ed.), *Kappa Distributions* (Theory and Applications in Plasmas. Elsevier, Amsterdam, Netherlands, 2017)

46. D. Summers, R.M. Thorne, A new tool for analyzing microinstabilities in space plasmas modeled by a generalized Lorentzian (Kappa) distribution. J. Geophys. Res.: Space Phys., 97, 16827–16832 (1992). https://doi.org/10.1029/92JA01664

47. M.A. Hellberg, R.L. Mace, Generalized plasma dispersion function for a plasma with a kappa-Maxwellian velocity distribution. Phys, Plasmas **9**, 1495–1504 (2002). https://doi.org/10.1063/1.1462636

48. I.B. Bernstein, Waves in a plasma in a magnetic field. Phys. Rev. **109**, 10–21 (1958). https://doi.org/10.1103/PhysRev.109.10

49. M.A. Milla, E. Kudeki, Incoherent scatter spectral theories-part II: modeling the spectrum for modes propagating perpendicular to *b*. IEEE Trans. Geosci. and Remote Sensing **49**, 329–345 (2011). https://doi.org/10.1109/TGRS.2010.2057253

50. J.-P. St-Maurice, On a mechanism for the formation of VLF electrostatic emissions in the high latitude F region. Planet. Space Sci. **26**, 801–816 (1978). https://doi.org/10.1016/0032-0633(78)90104-6

51. W.J. Miloch, H.L. Pécseli, J.K. Trulsen, Unstable ring-shaped ion distribution functions induced by charge-exchange collisions. Plasma Phys. Control. Fusion **55**, 124006 (2013). https://doi.org/10.1088/0741-3335/55/12/124006

52. S. Chandrasekhar. Plasma Physics. The University of Chicago Press, Chicago, Notes compiled by S.K. Trehan after a course given by S. Chandrasekhar (1960)

53. L.D. Landau, L.P. Pitaevskii, E.M. Lifshitz, *Electrodynamics of Continuous Media*, Course of Theoretical Physics, vol. 8, 2nd edn. (Butterworth-Heinemann, Oxford, UK, 1984)

54. T. O'Neil, Collisionless damping of non linear plasma oscillations. Phys. Fluids **8**, 2255–2262 (1965). https://doi.org/10.1063/1.1761193

55. V.I. Karpman, E.M. Maslov, Perturbation-theory for solitons. Sov. Phys. JETP, 46:281–291, Russian original Zh. Eksp. Teor. Fiz. **1977**(73), 537–559 (1977)

56. V.I. Karpman, J.P. Lynov, P. Michelsen, H.L. Pécseli, J. Juul Rasmussen, V.A. Turikov, Modification of plasma solitons by resonant particles. Phys. Rev. Lett. 43, 210–214 (1979). https://doi.org/10.1103/PhysRevLett.43.210

57. R.W. Motley, *Q Machines* (Academic Press, New York, 1975)

58. R. Schunk, A. Nagy, *Ionospheres: Physics, Plasma Physics, and Chemistry*, 2nd edn. (Cambridge University Press, UK, 2009). doi:https://doi.org/10.1017/CBO9780511635342

59. A.V. Gurevich, *Nonlinear Phenomena in the Ionosphere, Physics and Chemistry in Space*, vol. 10 (Springer, New York, 1978)

60. P.L. Bhatnagar, E.P. Gross, M. Krook, A model for collision processes in gases. I. small amplitude processes in charged and neutral one-component systems. Phys. Rev., 94, 511–525 (1954). https://doi.org/10.1103/PhysRev.94.511

61. N.A. Krall, A.W. Trivelpiece, Principles of Plasma Physics (McGraw-Hill. New York (1973). https://doi.org/10.1119/1.1987587

62. L. Stenflo, M.Y. Yu, P.K. Shukla, Shielding of a slow test charge in a collisional plasma. Phys. Fluids **16**, 450–452 (1973). https://doi.org/10.1063/1.1694361

63. V.L. Rekaa, H.L. Pécseli, J.K. Trulsen, Self-similar space-time evolution of an initial density discontinuity. Phys. Plasmas **20**, 072117 (2013). https://doi.org/10.1063/1.4816953

64. D.K.C. MacDonald, *Noise and Fluctuations: an Introduction* (John Wiley & Sons, New York, 1962)

65. H.B. Callen, T.A. Welton, Irreversibility and generalized noise. Phys. Rev. **83**, 34–40 (1951). https://doi.org/10.1103/PhysRev.83.34

66. B.B. Kadomtsev, *Plasma Turbulence* (Academic Press, New York, 1965)

67. T.H. Dupree, *Turbulence in Fluids and Plasmas*, volume XVIII of *Microwave Research Institute Symposia Series*, chapter 2 "Introduction to basic phenomena of turbulence in plasmas", pages 3–12. (Polytechnic Press, Brooklyn, N.Y., USA, 1969). ISBN: 0471274348, 9780471274346

68. P.L. Similon, R.N. Sudan, Plasma turbulence. Ann. Rev. Fluid Mech. **22**, 317–347 (1990). https://doi.org/10.1146/annurev.fl.22.010190.001533

69. S. Galtier, Wave turbulence in magnetized plasmas. Nonlin. Processes Geophys. **16**, 83–98 (2009). https://doi.org/10.5194/npg-16-83-2009

70. H.L. Pécseli, *Low frequency waves and turbulence in magnetized laboratory plasmas and in the ionosphere* (IOP Publishing, UK, 2016). 978-0-7503-1251-6

71. P. Guio, H.L. Pécseli, The impact of turbulence on the ionosphere and magnetosphere. Front. Astron. Space Sci. 7 (2021). https://doi.org/10.3389/fspas.2020.573746

72. W. Kollmann, *Navier-Stokes Turbulence: Theory and Analysis* (Springer, Switzerland, 2019)

73. H. Tennekes, J.L. Lumley, *A First Course in Turbulence* (The MIT press, Cambridge, Massachusetts, 1972). -13: 978-0262200196

74. D. Biskamp, *Magnetohydrodynamical Turbulence* (Cambridge University Press, Cambridge, 2003)

75. D. Biskamp, E. Schwarz, A. Zeiler, A. Celani, J.F. Drake, Electron magnetohydrodynamic turbulence. Phys. Plasmas **6**, 751–758 (1999). https://doi.org/10.1063/1.873312

76. M. Kono, H.L. Pécseli, Cascade conditions in electron magneto-hydrodynamic turbulence. Phys. Plasmas **29**, 122305 (2022). https://doi.org/10.1063/5.0124404

77. W. Horton, Ion acoustic turbulence and anomolous transport. J. Stat. Phys. **39**, 739–754 (1985). https://doi.org/10.1007/BF01008363

78. M. Kono, M.M. Škorić, *Nonlinear Physics of Plasmas*. Number 62 in Springer Series on Atomic, Optical, and Plasma Physics. Springer, Heidelberg, Germany, 2010

79. H. Ratcliffe, C.S. Brady, M.B. Che Rozenan, V.M. Nakariakov, A comparison of weak-turbulence and particle-in-cell simulations of weak electron-beam plasma interaction. Phys. Plasmas, 21, 122104 (2014). https://doi.org/10.1063/1.4904065

80. J.J. Rasmussen, K. Rypdal, Blow-up in nonlinear Schroedinger equations-I. A general review. Phys. Scripta **33**, 481–497 (1986). https://doi.org/10.1088/0031-8949/33/6/001

81. G. Sun, D.R. Nicholson, H.A. Rose, Statistical theory of cubic Langmuir turbulence. Phys. Fluids, 28 (1985). https://doi.org/10.1103/PhysRevLett.54.1664

82. P. Guio, F. Forme, Zakharov simulations of Langmuir turbulence: Effects on the ion-acoustic waves in incoherent scattering. Phys. Plasmas **13**, 122902 (2006). https://doi.org/10.1063/1.2402145

83. M.C. Kelley, *The Earth's Ionosphere, Plasma Physics and Electrodynamics*, volume 43 of *International Geophysics Series* (Academic Press, San Diego, California, 1989). ISBN-13: 978-0120884254

84. I.B. Bernstein, J.M. Greene, M.D. Kruskal, Exact nonlinear plasma oscillations. Phys. Rev. **108**, 546–550 (1957). https://doi.org/10.1103/PhysRev.108.546

85. H. Schamel, Electron holes, ion holes and double layers. Phys. Reports **140**, 161–191 (1986). https://doi.org/10.1016/0370-1573(86)90043-8

86. H. Bindslev, Dielectric effects on thomson scattering in a relativistic magnetized plasma. Plasma Phys. Controlled Fusion **33**, 1775–1808 (1991). https://doi.org/10.1088/0741-3335/33/14/005

87. J.P. Dougherty, D.T. Farley Jr. A theory of incoherent scattering of radio waves by a plasma: 3. scattering in a partly ionized gas. J. Geophys. Res. (1896–1977) 68, 5473–5486 (1963). https://doi.org/10.1029/JZ068i019p05473

88. T. Hagfors, R.A. Brockelman, A theory of collision dominated electron density fluctuations in a plasma with applications to incoherent scattering. Phys. Fluids **14**, 1143–1151 (1971). https://doi.org/10.1063/1.1693578

89. T. Hagfors, Note on the scattering of electromagnetic waves from charged dust particles in a plasma. J. Atmospheric Terrestrial Phys. **54**, 333–338 (1992). https://doi.org/10.1016/0021-9169(92)90012-A

90. F. Melandsø, T.K. Aslaksen, O. Havnes, A kinetic model for dust acoustic waves applied to planetary rings. J. Geophys. Res.: Space Phys. 98, 13315–13323 (1993). https://doi.org/10.1029/93JA00789

91. J.Y.N. Cho, C.M. Alcala, M.C. Kelley, W.E. Swartz, Further effects of charged aerosols on summer mesospheric radar scatter. J. Atmospheric Terr. Phys. **58**, 661–672 (1996). https://doi.org/10.1016/0021-9169(95)00065-8

92. C. La Hoz, Radar scattering from dusty plasmas. Phys. Scripta **45**, 529–534 (1992). https://doi.org/10.1088/0031-8949/45/5/021

93. S.C. Buchert, A.P. van Eyken, T. Ogawa, S. Watanabe, Naturally enhanced ion-acoustic lines seen with the EISCAT Svalbard Radar. Adv. Space Res. **23**, 1699–1704 (1999). https://doi.org/10.1016/S0273-1177(99)00382-8

94. M.T. Rietveld, B. Isham, T. Grydeland, C. La Hoz, T.B. Leyser, F. Honary, H. Ueda, M. Kosch, T. Hagfors, HF-pump-induced parametric instabilities in the auroral E-region. Adv. Space Res. **29**, 1363–1368 (2002). https://doi.org/10.1016/S0273-1177(02)00186-2

95. T. Grydeland, C. La Hoz, T. Hagfors, E.M. Blixt, S. Saito, A. Strømme, A. Brekke, Interferometric observations of filamentary structures associated with plasma instability in the auroral ionosphere. Geophys. Res. Letters **30**, 1338 (2003). https://doi.org/10.1029/2002GL016362

96. F. Sedgemore-Schulthess, J.-P. St-Maurice, Naturally enhanced ion-acoustic spectra and their interpretation. Surveys Geophys. **22**, 55–92 (2001). https://doi.org/10.1023/A:1010691026863

97. W.B. Davenport, W.L. Root, *An Introduction to the Theory of Random Signals and Noise* (McGraw-Hill, New York, 1958)

98. R.K. Pathria, *Statistical Mechanics*, 2nd edn. (Butterworth-Heinemann, Oxford, 1996). 0 7506 2469 8

99. M.J. Lighthill, *An Introduction to Fourier Analysis and Generalized Functions* (Cambridge University Press, London, 1964)

100. E. Berz. *Verallgemeinerte Funktionen und Operatoren*, volume 122/122a of *Hochschultaschenbücher*. Bibliographisches Institut, Mannheim, 1967

101. A.S. Davydov, *Quantum Mechanics* (Pergamon Press, Oxford, 1965)

102. A.I. Khinchine, *Mathematical Foundations of Statistical Mechanics* (Dover, New York, 1949)

103. J.S. Bendat, Principles and Applications of Random Noise Theory (John-Wiley & Sons. New York (1958). https://doi.org/10.1002/qj.49708536318

104. P. Claps, A. Giordano, F. Laio, Advances in shot noise modeling of daily streamflows. Adv. Water Resources **28**, 992–1000 (2005). https://doi.org/10.1016/j.advwatres.2005.03.008

105. A. Theodorsen, O.E. Garcia, Level crossings, excess times, and transient plasma-wall interactions in fusion plasmas. Phys. Plasmas **23**, 040702 (2016). https://doi.org/10.1063/1.4947235

106. S. O. Rice. Mathematical analysis of random noise, I. *Bell System Tech. J.*, 23:282–332, 1944. reprinted by Wax, N. in: Selected Papers on Noise and Stochastic Processes (1954) Dover, New York

107. S. O. Rice. Mathematical analysis of random noise, II. *Bell System Tech. J.*, 24:46–156, 1945. reprinted by Wax, N. in: Selected Papers on Noise and Stochastic Processes (1954) Dover, New York

108. I. Bar-David, A. Nemirovsky, Level crossings of nondifferentiable shot processes. Information Theory, IEEE Transactions on **18**(1), 27–34 (1972). https://doi.org/10.1109/TIT.1972.1054748

109. N. Campbell. The study of discontinuous phenomena. *Proc. Phil. Soc.*, 15:117–136, 1909. and ibid. Discontinuities in light emission, 310-328

110. E.N. Rowland, The theory of the mean square variation of a function formed by adding known functions with random phases, and applications to the theories of the shot effect and of light. Math. Proc. Cambridge Phil. Soc. 32, 580–597 12 (1936). https://doi.org/10.1017/S0305004100019319

111. M.A. Lieberman, A.J. Lichtenberg, *Principles of Plasma Discharges and Materials Processing*, 2nd edn. (John Wiley & Sons Inc, New York, 1994)

112. M. Kono, H.L. Pécseli, Dynamic evolutions of Bohm sheaths and pre-sheaths. Phys. Plasmas **31**, 023508 (2024). https://doi.org/10.1063/5.0176287

113. C.-L. Wang, G. Joyce, D.R. Nicholson, Debye shielding of a moving test charge in plasma. J. Plasma Phys. **25**(2), 225–231 (1981). https://doi.org/10.1017/S0022377800023084

114. T. Birmingham, J. Dawson, C. Oberman, Radiation processes in plasmas. Phys. Fluids **8**, 297–307 (1965). https://doi.org/10.1063/1.1761223

115. L. Spitzer, *Physics of Fully Ionized Gases, Interscience Tracts on Physics and Astronomy*, vol. 3 (Interscience, New York, 1965)

116. J.A. Stratton, *Electromagnetic Theory* (McGraw-Hill Book Company, 1941)

117. J.D. Jackson, *Classical Electrodynamics*, 2nd edn. (John Wiley & Sons, New York, 1975)

118. L.A. Vaĭnshteĭn, Propagation of pulses. Usp. Fiz. Nauk., 118:339–367, See also (1976) Sov. Phys. Usp. **19**, 189–205 (1976)

119. D. Jovanović, H.L. Pécseli, K. Thomsen, Non-linear transient signal propagation in homogeneous plasma. J. Plasma Phys. **28**, 159–175 (1982). https://doi.org/10.1017/S0022377800000167

120. P. Guio, W.J. Miloch, H.L. Pécseli, J. Trulsen, Patterns of sound radiation behind pointlike charged obstacles in plasma flows. Phys. Rev. E **78**, 016401 (2008). https://doi.org/10.1103/PhysRevE.78.016401

121. J.R. Sanmartin, S.H. Lam, Far-wake structure in rarefied plasma flows past charged bodies. Phys. Fluids **14**, 62–71 (1971). https://doi.org/10.1063/1.1693289

122. T. Huld, H. L. Pécseli, and J. Juul Rasmussen. Ion-acoustic wave propagation in plasma with ion beams having a finite cross section. *IEEE Trans. Plasma Sci.*, 18:149–158, 1990. doi:https://doi.org/10.1109/27.45518

123. P. Guio, H.L. Pécseli, Radiation of sound from a charged dust particle moving at high velocity. Phys. Plasmas **10**, 2667–2676 (2003). https://doi.org/10.1063/1.1585031

124. R.J. Mason, Structure of evolving ion-acoustic fronts in collisionless plasmas. Phys. Fluids **13**, 1042–1048 (1970). https://doi.org/10.1063/1.1693006

125. S.A. Andersen, G.B. Christoffersen, V.O. Jensen, P. Michelsen, P. Nielsen, Measurements of wave-particle interaction in a single ended Q-machine. Phys. Fluids **14**, 990–998 (1971). https://doi.org/10.1063/1.1693560

126. P. Michelsen, H.L. Pécseli, Propagation of density perturbations in a collisionless Q-machine plasma. Phys. Fluids **16**, 221–225 (1973). https://doi.org/10.1063/1.1694322

127. H.W. Ellis, R.Y. Pai, E.W. McDaniel, E.A. Mason, L.A. Viehland, Transport properties of gaseous ions over a wide energy range. Atomic Data and Nuclear Data Tables **17**, 177–210 (1976). https://doi.org/10.1016/0092-640X(76)90001-2

128. J.B. Hasted, Atom-atom interchange collisions by drifting ions. Proc. Phys. Soc. **86**, 795 (1965). https://doi.org/10.1088/0370-1328/86/4/316

129. B.A. Trubnikov, Particle interactions in fully ionized plasmas, in *Reviews of Plasma Physics*. ed. by M.A. Leontovich. volume 1. (Consultants Bureau, New York, 1965), pp.105–204

Index

© The Editor(s) (if applicable) and The Author(s), under exclusive license
to Springer Nature Switzerland AG 2025
H. L. Pécseli, *Introduction to the Theory of Incoherent Scattering of Radar Waves from
Plasmas*, SpringerBriefs in Physics, https://doi.org/10.1007/978-3-031-82652-8